Women, Prisons and Psychiatry:
mental disorder behind bars

Tony Maden
Senior Lecturer, Department of Forensic Psychiatry,
Institute of Psychiatry, London, UK

BUTTERWORTH
HEINEMANN

Butterworth-Heinemann Ltd
Linacre House, Jordan Hill, Oxford OX2 8DP

℞ A member of the Reed Elsevier plc group

OXFORD LONDON BOSTON
MUNICH NEW DELHI SINGAPORE SYDNEY
TOKYO TORONTO WELLINGTON

First published 1996

© Butterworth-Heinemann Ltd 1996

British Library Cataloguing in Publication Data
A catalogue record for this book is available from the
British Library

Library of Congress Cataloguing in Publication Data
A catalogue record for this book is available from the
Library of Congress

ISBN 0 7506 2003 X

Typeset by BC Typesetting, Bristol
Printed and bound in Great Britain by Biddles Ltd, Guildford and King's Lynn

Women, Prisons and Psychiatry:
mental disorder behind bars

Contents

Preface

This book should be of interest to anyone who wants to know more about psychiatric disorder in prison, but I hope it will be of most use to those who are able to do something about the problem.

The core of the book is a study of psychiatric disorder in women serving a prison sentence in England and Wales. It was carried out as part of a larger project by the Department of Forensic Psychiatry at the Institute of Psychiatry, which included a sample of male prisoners. The methods used were the same for the women and the men, allowing straightforward comparisons.

The methods used in the survey were scientific, as far as possible, given the current limitations on psychiatry as a scientific discipline. Sampling of the prison population was carried out in order to ensure that the findings were representative. Psychiatric disorders were defined according to accepted criteria and standards. The decisions of the interviewers were overseen by a research panel, in an attempt to improve reliability and validity. The outcomes of the study include measures of how much psychiatric disorder is present in the female prison population. The figures are not meant to be the absolute truth, but represent the best current estimate, using accepted epidemiological standards.

However, in one important respect, both the study and the book departed from a rigid scientific approach. They are based on the premise that women (and men) with a serious psychiatric disorder should not be in prison. Therefore the description of mental disorder in prison leads automatically into a discussion of what can be done to remedy an unacceptable situation. Some of the recommendations are general and require large-scale changes. Other recommendations are more easily implemented, and could inform the practice of anyone working with offenders and/or the mentally disordered.

It follows that the book should be judged in the long term by the extent to which it improves the lot of the mentally disordered offenders.

T M

Acknowledgements

The survey at the heart of this book was undertaken while the author was engaged in preparing a psychiatric profile of the prison population, under the supervision of Professor John Gunn and in cooperation with Dr Mark Swinton, in the Department of Forensic Psychiatry at the Institute of Psychiatry. The project was funded by the Directorate of the Prison Medical Service at the Home Office and the Director, Dr Rosemary Wool, provided valuable encouragement and advice, as did many other members of the DPMS and the Home Office Research and Planning Unit. Thanks are due to all medical officers and prison staff who were interviewed during the study. Particular mention must be made of Dr Dorothy Speed, then principal medical officer at HMP Holloway. She can take credit for many improvements to medical services at Holloway, but also provides an excellent example of the way in which a doctor can exert a positive influence on the general regime of a prison.

The author wishes to thank Professor John Gunn and Dr Mark Swinton, my colleagues on the prison survey. The research panel for the survey also included Dr Pamela Taylor, Dr David Tidmarsh, Dr Christine Curle, Mr Angus Cameron and Mr Graham Allison. Dr Graham Robertson deserves particular thanks, as a member of the research panel who also commented on earlier drafts. Although not directly involved in the survey, Drs Paul Bowden and James MacKeith have also been influential, in the way they practise forensic psychiatry. Mrs Maureen Bartholomew and Mrs Carol Double provided secretarial and organizational assistance.

The project could not have been carried out without the co-operation of a large number of women and men serving a prison sentence. They took part without any prospect of direct benefit, often at some personal cost, discussing painful matters with a stranger. Many expressed a wish that their cooperation would lead to an improvement in medical services for prisoners, and I hope they will not be disappointed.

Parts of this book were included in the author's thesis for the degree of Doctor of Medicine at the University of London (Maden, 1992).

1
Women and psychiatrists in prisons

Introduction

This book presents a psychiatrist's view of women in prison. It is based on a survey of psychiatric disorder in women serving a prison sentence, with an emphasis on the provision of psychiatric services. It provides an estimate of the number of female prisoners who have psychiatric problems. It describes the different types of psychiatric disorder from which they suffer and attempts to find connections between the disorder and offending. Finally, and most importantly, it asks what services should be provided in order to improve the lives of the women concerned.

The psychiatrist's perspective is a narrow one, and the book does not pretend to be a comprehensive account of women's offending or women's imprisonment. A medical view is no substitute for historical, feminist or other accounts, but it is valid and important in its own right. One theme of the book will be that psychiatry cannot provide wide-ranging explanations of crime, and doctors should confine their efforts to detecting and treating mental disorder in offenders. There is plenty of room for improvement in the way that health services carry out this task, before doctors start to stray into territory which is best left to the criminologist or the moral philosopher.

Despite this narrow perspective, psychiatry in prisons must be seen within a wider context. This chapter supplies that context. First, it asks why women are sent to prison. A few statistics on gender differences in offending are presented, before examining theoretical explanations. One of these explanations centres on mental disorder, leading on to the question of why psychiatrists go into prisons. The history of the redevelopment of HMP Holloway shows how psychiatry's role has changed in recent years, so that expectations are now more realistic. Finally, there is a discussion of the findings of previous psychiatric surveys of women in prison. They are shown to be rather limited in scope and provide no justification for assumptions that have been made in the past about the role of psychiatric disorder in women's offending.

In summary, this chapter attempts to answer two basic questions:

1. Why are women sent to prison?
2. Why do psychiatrists go into prisons?

Why are women sent to prison?

This question can be answered in two ways. First, there are the statistics on offending by women. Second, there are various attempts to place these offences within a theoretical framework. There is now a considerable literature on both topics and only a brief outline is presented here, with appropriate references for those who require a fuller account.

The statistics

The difference between male and female offending rates has been described as the most significant feature of recorded crime. In England and Wales in 1989, a total of 396 500 convictions or cautions for indictable offences were recorded against men compared to 76 200 for women, a ratio of 5:1 (Home Office, 1990a). Within these figures, violence against the person is even less common in women, accounting for 10 per cent of their indictable crime compared to 16 per cent in men. Property offences account for most crime in both sexes and male:female ratios for different types of crime are as follows: theft 3:1; violence against the person 8:1; drug offences 9:1; criminal damage 11:1; and sexual offences 106:1.

A similar pattern is found in the prison population, although the gender differences are now magnified. Most offenders are never sent to prison and men are much more likely than women to find themselves serving a sentence. In England and Wales, the female:male ratio for prisoners is roughly 1:30. This is a greater difference than is found for offending rates (with the exception of sexual offending). The gender difference is even greater in prisoners sentenced for violent offences, the female:male ratio being 1:39 (Home Office, 1989a).

One limitation of these figures is that they reveal little about qualitative differences in the offending of men and women. A constant problem in gender comparisons is that women tend to commit different offences from men, as well as offending less often. For example, the victims of women who commit murder are more likely to be family members or intimate friends than is the case for men (Gibson and Klein, 1969; Gibson, 1975). For every woman serving a sentence for homicide or attempted homicide there were 27 men, while the ratio among those sentenced for other violent offences was 1:53. It is not easy to explain this difference, as it contradicts the general principle that gender differences are greater for more serious offences. In a later chapter, women sentenced for murder will be described in more detail. For now, these figures serve as a reminder that simple offending rates may give a misleading picture of gender differences. The same offence label can represent quite different behaviours in men and women.

Offending is embedded in the life of the offender. Gender influences all aspects of life, so it is not surprising that differences in criminal behaviour are complex. Anyone tempted by simple or biological explanations should note that the proportion of women who were judges in 1989 was roughly equal to the proportion who were sentenced prisoners. In the top ranks of the civil service, the representation of women was even lower than within prison.

If offending rates reflect wider social differences, then changes in the role of women should be reflected in their offending. It has been argued that the greater participation of women in areas of society from which they were once excluded has led to a recent increase in female offending in general, and violent offending in particular. Box and Hale (1983) attempt a critical examination of this question and point to the considerable methodological problems involved – problems which have not been surmounted by most studies in the area. Their own study found no relationship between four measures of female liberation and female crime rates during the years 1951 to 1979. Jones (1991) argues that fears of a rising tide of female lawlessness, based on little evidence, have surfaced at various points in history but reflect male insecurity about changing gender roles rather than a deterioration in female standards of conduct.

The findings of Box and Hale show that gender differences cannot be considered in isolation from other influences on offending. It may be that changes in other social factors have had greater impact and have swamped any effect of changing gender roles. Biological factors may also play a role (it is difficult to think of any social explanation for the enormous gender difference in rates of sexual offending). The following section considers several attempts to explain gender differences.

Theoretical frameworks

The causes of gender differences in offending have been the subject of much argument, often amounting to little more than a repetition of the nature–nurture debate. A detailed account is beyond the scope of this book but Morris (1987) reviews the major theories and includes an overview of feminist critiques. Mandaraka-Sheppard (1986) places the different theoretical frameworks in an historical context. The present review considers four broad approaches. It is important to note that they are not mutually exclusive. One of the problems in this area is a tendency to put forward a single mechanism as the sole explanation for all gender differences observed within the criminal justice system. This is plainly absurd. The outcome of prosecutions depends on individuals, whether they are defendants, police, doctors, lawyers, judges or jurors. Different individuals may be influenced by different principles. The same person may be influenced by different principles in different cases and at different times. Therefore, no attempt is made to decide which of the following frameworks is 'correct', although some of the evidence for and against each is noted.

Chivalry and under-reporting
Recorded crime is only a small part of the total within any society, and official figures may give a misleading picture of female offending. Pollak (1950) went so far as to claim that gender differences in crime were an artefact, and that women committed just as much crime as men but it was less likely to be detected or reported. The examples given by Pollak included crime within the home, and theft by prostitutes from their clients. He also suggested that the male-dominated legal system operated according to principles of chivalry, so that women's crime, even when it was detected, was less likely to result in prosecution, conviction or imprisonment.

When Pollak put forward these ideas, little thought had been given to the problem of studying unrecorded crime. It was assumed that crime statistics were an accurate reflection of reality. His ideas were helpful in drawing attention to the fact that official statistics are a poor approximation of the real world.

There is now an extensive literature on 'unofficial' crime and the steps which come between a criminal act and eventual imprisonment. It is recognized that many factors, including gender, can operate at each stage to bias the criminal process. Although this work has shown that most crime goes undetected, there is no real support for Pollak's suggestion that gender differences would disappear when unrecorded crime was added to the picture. The non-accidental injury of children is a good example of an offence often committed by women and not acknowledged by professionals until the 1960s (Kempe et al., 1962), but there are many more examples of 'invisible' offending by men within the home, with their partners or their children as victims.

The British Crime Survey (Mayhew, Elliott and Dowds, 1989) confirmed the importance of under-reporting and showed that less serious offences are more likely to go unrecorded. The offending of women tends to be less serious than that of men, so it is possible that gender differences for minor offences would be decreased if unrecorded crime were taken into account. However, there is no evidence to suggest that the broad pattern of gender differences would be altered, and under-reporting becomes less important for serious offences.

In the same way, there is little evidence to support Pollak's contention that chivalry is a major or consistent influence on the treatment of women within the criminal justice system. D'Orban's (1971) review of the literature concluded that gender has an influence on the treatment of offenders at all stages of the criminal justice system but the effects are complex. He was unable to identify an overall trend in either direction and it is possible that gender has different effects in relation to different offences. Feminism has emphasized the possibility that, far from being treated better, women may be placed at a disadvantage in the male-dominated criminal justice system. This assertion will now be explored, before considering further evidence.

The gender politics of women's imprisonment
It has been argued that women are punished, including being sent to prison, for failing to comply with socially-conditioned female stereotypes. In other words, women are sent to prison because they break the rules governing female behaviour, rather than the laws in the statute books. The implication is that women are getting a raw deal. They face a form of double jeopardy and must conform to a restrictive gender stereotype, or risk being criminalized.

Most judges are male and it is not difficult to find statements and decisions which support this thesis. However, the evidence is not all one-sided, and it is also possible to find examples of apparent leniency by courts towards female defendants. The quoting of examples, for and against, misses the point. The politics of gender are important in the operation of the courts, but other factors may be of equal or greater weight in a particular case. If one considers only gender issues, the picture can be as biased and incomplete as when gender was ignored.

This point is illustrated by recent developments in the study of gender and violence. Feminism has given a new impetus to this topic, with attempts to relate violence by women to their role in society. Hartmann (1977) describes thirteen middle-class Victorian women found guilty of murder and suggests that their acts of violence were extreme solutions to problems faced by many of their contemporaries. Jones (1991) adopts a similar approach, in an ambitious attempt to relate changes in the nature of female homicide to the history of women in America.

The value of these books is that they abandon simple, deterministic accounts and demonstrate the futility of explaining gender differences in violent offending without reference to the wider context of women's role in society. Their weakness is that, by emphasizing the common ground between women who kill and the majority who do not, suggesting that female murderers are 'just like other women', it is easy to lose sight of those factors which do distinguish this tiny minority.

A recent and comprehensive review of research and statistics provides little support for the feminist view of the criminal justice system as a whole. The authors asked the ambitious question: 'Does the criminal justice system treat men and women differently?' (Hedderman and Hough, 1994). Their thorough and critical account concluded that women were not systematically dealt with more severely than men (although there may, of course, be individual cases of women who receive unusually harsh punishment). In fact, the available figures suggested that a higher proportion of women are cautioned for serious offences, women are less likely than men to be remanded in custody and, overall, women seem more likely to receive lenient sentences. This last finding remained valid, even when previous convictions were taken into account. In addition, although the authors confirmed that women were less likely to commit crimes than men, self-report studies showed that official statistics exaggerate the difference, implying that women's offending was more likely to escape detection, prosecution or conviction. Perhaps chivalry is not yet dead.

Just deserts?

'You do the crime, you do the time.'
Carol, HMP Holloway, 1989

Carol (not her real name) was in Holloway serving four years for the possession of cocaine with intent to supply. A middle-aged woman from South London, with a family, she sold cocaine for a living. After a few years, she had been arrested. She had been caught before, and expected to be caught again in the future. As a woman who relied on criminal activity for her livelihood, she saw imprisonment as a form of taxation. Like many business people, she did not relish paying, but neither did she question the right of the state to collect its dues.

Although Carol's views may be oversimplified (and amoral), they are a fair reflection of the 'just deserts' model of punishment. Offenders are meant to receive a penalty which is proportional to the seriousness of their crime. They pay their debt to society, and are free to go. There is no mention here of unfairness, or punishment for failure to comply with stereotypes. There is no mention either of rehabilitation or reform. Neither Carol nor the prison intended that she should be changed or made into a better citizen. She would have resisted any such attempt as intrusive and unjustified. In the same way, she did not see herself as a victim, whether of gender bias or other aspects of the criminal justice system. Carol's view was that she had made her choices, and was prepared to live with the consequences.

Of course, it would be complacent to believe that all offenders appearing before the courts receive their just deserts. The recent spate of miscarriages of justice reminds us that innocent people may be punished and the guilty go free. Gender, race, social class and many other factors are likely to influence judges and juries, often in subtle and complex ways of which they may be unaware. Nevertheless, the criminal justice system in the UK has not yet reached the stage where the concept of just deserts has no meaning.

The medical model and 'psychiatrization'

> Most women and girls in custody require some form of medical, psychiatric or remedial treatment. The main feature of the programme is . . . the building of an establishment that is basically a secure hospital to act as the hub of the female prison system. Its medical and psychiatric facilities will be its central feature, and normal custodial facilities will comprise a relatively small part of the establishment.
>
> James Callaghan (1968)

This was part of the Secretary of State's announcement of plans for the redevelopment of Holloway prison. It reflects the spirit of the times, when there was general optimism about rehabilitation and the role that could be played by psychiatry and psychology. Typical of the era are books such as *The Crime of Punishment* (Menninger, 1969) in which it was argued that to punish anyone was both wrong and counter-productive, when the real need was for intervention aimed at modifying future behaviour. In this model, offending was stripped of its moral dimension and became a technical problem for the behavioural sciences and medicine. Sim (1990) documents the history of these ideas, particularly in relation to female prisoners, whom he characterizes as being 'at the centre of the professional gaze' for over 150 years. The process of relabelling female deviance as mental disorder rather than criminality has also been called 'medicalization' or 'psychiatrization'. By shifting transgression from a moral to a medical field, the appropriate response becomes treatment rather than punishment. Sim argues that this is not a benign process, as both treatment and punishment are means of social control, by which the state maintains its power over those who would break its laws.

The depiction of psychiatry as an agent of social control is developed further in the work of Foucault. For the present, it is sufficient to note that the notion of 'curing' female offenders did not originate in the 1960s, although it does seem to have reached its peak around this time. After all, the truly remarkable thing about Callaghan's statement is not the attitudes it implied, but the fact that it was soon to be implemented as the central plank of penal policy for women in the United Kingdom. This supreme example of 'psychiatrization' warrants further consideration.

The redevelopment of HMP Holloway

The old Holloway prison was a standard Victorian design, with tiered galleries of cells on wings radiating from a central point. The new Holloway looks more like a modern university. It sprawls across the site, with long and disorienting corridors joining the different units. The plans included no 'hospital wing', which was considered unnecessary in view of the therapeutic nature of the whole place. The first governor of the prison was to be a doctor, reflecting the mixing of medical and disciplinary functions. Therapeutic optimism even went as far as discussions of a future when most female offenders had been treated or cured. Holloway would no longer be needed as a prison, and the building could be handed over to the health service for incorporation into its system of secure hospitals.

The experiment was a failure. Rather than a model for other prisons, the new Holloway became notorious for its repressive regime. Assaults on staff and inmates were a daily occurrence and dormitories were put out of use by vandalism at the rate of one per month. Self-mutilation was common and a minority of seriously mentally ill women were held in unacceptable and squalid conditions (Moorehead, 1985).

A central problem was that doctors no longer acted as gatekeepers for the psychiatric services. The traditional job of the prison doctor, when rules are broken, is to decide on fitness for punishment. In practice, this usually means a decision as to who is mentally disordered and who is not. The presence of serious mental disorder implies a need for treatment rather than punishment and vice versa. Once doctors abandon this role, seriously mentally ill women are punished for their actions. They lose privileges and are held in worse conditions, leading to a further deterioration in mental state and behaviour, to which the response is further punishment. If this vicious circle is not interrupted, it leads to inhumane conditions where self-harm, suicide and violence are common.

A further problem at Holloway was that women who were not mentally disordered believed they were treated as if they were not responsible for their actions. It was as if the prison had decided to make all its inmates into patients, without asking the women themselves. It is hardly surprising that many resented this assumption.

After a few difficult years, the Home Office concluded that most women in prison were 'depressingly normal', in that their offending was not a result of psychiatric disorder and was not amenable to medical treatment. Holloway was reorganized to separate disciplinary and medical systems. Doctors identify and treat the sick, arranging their transfer to outside hospitals when necessary. The governor is responsible for discipline.

These changes have brought Holloway back into line with the male prison system. Both now operate according to the principle of 'humane containment'. This states that the function of prisons is simply to hold inmates in humane conditions for as long as is specified by the courts. The punishment imposed by the court is loss of liberty and the prisons should not impose further punishment in the form of arduous conditions. On the negative side, humane containment also implies that rehabilitation is not a priority. Earlier optimism has now given way to extreme pessimism, the assumption being that prison makes many people worse and there is no point in striving to make it a place where people can be helped.

With hindsight, the attempt to make a prison into a hospital was misguided and doomed to failure. On the other hand, it is possible to go too far in a backlash against rehabilitation, so that help is not available to those women who need it and could benefit from it. There will be further discussion of this point in the final chapter, when the extent of psychiatric disorder has been described.

The experiment at Holloway raises questions about the role of psychiatrists within prison and this issue will now be examined in more detail.

Why do psychiatrists go into prisons?

The simple answer to this question should be: 'in order to treat mental disorder'. The reality is more complicated, mainly because of the way in which medical and disciplinary functions can easily become confused. Doctors in prison have always faced a conflict of loyalties, with the demands of the institution set against the needs of the patient/prisoner. When the institution is also the doctor's employer, it cannot be assumed that the patient's needs will always be given priority.

Psychiatry and discipline

The events at Holloway provide an interesting perspective on the role of doctors in prison discipline. When a woman is charged with an offence against prison discipline, a doctor is required to certify that she is fit for adjudication. If she is pronounced fit, the charge will be heard and, if it is proven, punishment will be given. It is easy to understand the ethical concerns surrounding doctors' involvement in the disciplinary process. At first sight, it appears to confirm all the criticisms made by Foucault and Sim, that psychiatry is nothing more than a tool for maintaining order.

However, there is a more benign aspect to doctors' involvement in this process. When mental disorder is detected, adjudication and punishment are avoided in favour of treatment. Their function is the same as that of the psychiatrist in the magistrate's court, attempting to divert mentally ill offenders into treatment facilities. The lesson of Holloway is that someone has to do this job. Otherwise, punishment is given to mentally ill women who are not fully responsible for their actions and an institution can quickly descend into chaos. Directing the mentally ill towards treatment rather than punishment is the acceptable face of psychiatrization. If carried out properly, it is an extension of the basic function of treating mental disorder.

A cure for crime?

Both mental disorder and crime can be seen as forms of deviance. This is a useful perspective for the sociologist but causes enormous problems if carelessly adopted by psychiatrists. The temptation is to assume that expertise in one area will lead to success in the other. Two problems follow. The first is unfounded optimism about treating offenders and preventing reoffending. The second problem, when psychiatry fails to deliver on this rash promise, is a backlash of therapeutic nihilism. The prison psychiatrist opts out and claims that nothing can be done for offenders, even when there is evidence of psychiatric disorder.

The history of psychiatry in prison has seen a shift from the optimistic view to the pessimistic. In part, this is a reflection of more general changes within criminology. Early explanations of crime focused on the characteristics of the offender, beginning with Lombroso's documentation of physical characteristics and developing into a quest to identify the psychological features which distinguish offenders from non-offenders. Psychiatry was central to this search for individual causes of offending but success was minimal. Since the 1960s, there has been much greater emphasis on the social causes of crime and the relative nature of deviance. The concept of labelling was discredited. Behaviour was no longer to be explained in terms of individual differences, but as the product of the political and social forces to which an individual was subjected. This emphasis was strengthened by the anti-psychiatry movement of Laing and Szasz, who refused to acknowledge the reality of psychiatric disorder.

The redevelopment of Holloway was the peak of psychiatric intervention in women's offending. Since then, there has been something of a backlash, and it will be argued that we have now reached a point where too little help is available for those offenders who need it. Prison staff have become demoralized. The phrase 'humane containment' may have a pleasant ring to it but it amounts to little more than 'lock 'em up and count 'em'. In the absence of other aims, it is difficult to avoid cynicism and disillusionment. This has been recognized by the prison service in its recent statement of purpose, which sets out cautiously ambitious goals.

For mentally disordered prisoners, there has also been a recent shift. This is reflected in the Reed report on mentally disordered offenders (Department of Health and Home Office, 1992). Reed's general approach avoids extremes of optimism or pessimism, and portrays offenders as a sub-section of the general population, with a high prevalence of psychiatric disorder. Mentally disordered offenders have special needs for health and social services, and the challenge is to deliver these services as efficiently as possible. The aim is that the mentally disordered are treated, without being disadvantaged by the fact that they are offenders. Psychiatry is removed from issues of morality and offending, and reinstated as a branch of medicine. Physical illness provides a good analogy for this approach. If a prisoner develops appendicitis or suffers a heart attack, she or he is moved to hospital for treatment. Priority is given to safe and appropriate treatment of the medical condition. Moral questions do not arise; it would be ridiculous to ask whether treatment affects the future chances of reoffending. The question is more complex for psychiatric disorders, where it is reasonable to expect that treatment may reduce the chances of reoffending. Nevertheless, the basic principle of Reed is that mental disorder should be treated for its own sake, because it is there, not because of presumed effects on future offending.

This approach has the advantage of aligning psychiatry with the rest of medicine. The doctor's role is much clearer. The nature and extent of psychiatric disorder are defined, and the appropriate range of services is organized in response. The study described in the following chapters adopted this stance. The first aim was to measure the extent of psychiatric disorder, then to estimate the need for treatment. Such an approach seems straightforward and self-evident, but it is only in recent years that prison surveys have attempted to measure the need for services. This begs the question, what was the purpose of earlier prison studies? The following chapter attempts an answer, by reviewing the aims and findings of previous psychiatric surveys of women in prison.

2

Previous surveys of women in prison

Introduction

This chapter is concerned with what psychiatrists have found when they have studied women in prison. Like most people, psychiatrists tend to find the things they are looking for. One of the themes of this chapter is that early researchers were looking mainly for causes of crime, having less interest in service provision and, therefore, in determining accurate prevalence rates of psychiatric disorders.

In describing the results of previous surveys, the emphasis is on psychiatrists as researchers rather than practitioners. In fact, there is considerable overlap, and most studies were done by doctors working in the field, usually as prison medical officers. Another theme of the chapter will be the way in which this has influenced the nature of the studies. Both the strengths and weaknesses of early studies can be related to the fact that they were carried out by prison doctors. However, it is paradoxical that these studies, conducted by people providing a service, made little reference to service provision. Instead, the impression is of practitioners who had the curiosity and the time to step back from their everyday work, and reflect on the causes and meaning of deviance. In today's prisons, life is more pressured, and most of that pressure arises from the difficulty of cajoling or persuading the health service (i.e. the NHS) to respond to prisoners' needs. This change in the position of prison doctors says much about wider changes in the provision of psychiatric services, and the impact of hospital bed closures.

It is usual to preface any discussion of women and offending with the qualification that little work has been done in the area. In general terms, this is no longer true. With the growth of feminism, there has been a rapid expansion in the literature of women and all aspects of the criminal justice system, including prison. Most of this literature is sociological. It attempts to describe women's experience, and the forces which shape that experience. By contrast, empirical studies of psychiatric disorder in female offenders or prisoners remain rare. In addition, all the existing studies of mental disorder in female prisoners have important limitations, which prevent a valid comparison with studies of men in prison.

Problems in prison research

Before describing previous findings, it is worth considering some of the reasons for the lack of studies, and their limitations. Of course, there are general and obvious reasons, including the neglect of women by criminology, a situation which is now being rectified. There is also a dearth of research in forensic psychiatry, which has developed as a clinical rather than an academic discipline. However, in addition to these general problems, there are more specific reasons for the lack of psychiatric research in prisons. Some are practical, others are theoretical or ideological.

Practical problems

Working within prisons is difficult. Access requires the cooperation of the prison department and may involve bureaucratic delays. Security considerations place further limits on access. Visitors must be escorted. The prison regime means that inmates are available for short periods only, and the short working day adds to the cost of research. Prisons are often busy and poorly resourced, without the facilities to assist researchers.

As a consequence of these difficulties, much of the work has been done by prison staff, rather than outside researchers. It is inevitable that prison staff will have different priorities. The understandable temptation for a prison doctor is to survey inmates within one prison, who would be seen anyway in the course of his or her work. The sampling is based on convenience, rather than being designed to address a specific question. When questions are asked, they may not be the questions that interest researchers outside prison. For example, a lot of prison research has failed to look beyond the prison walls at psychiatric provision in the community.

Inmates and staff are often suspicious of research, if not hostile. Staff may simply give research a low place on their list of priorities, when they are already hard pressed. Others are more actively hostile, and oppose any perceived attempt to help or be 'soft' on inmates. Psychiatry also generates specific resistance. Most inmates would rather be regarded as bad than mad, and fear being 'nutted off' (prison slang for transfer to a psychiatric hospital). A study must be carefully explained and introduced if inmates are to be persuaded to take part, and this requires a high level of staff cooperation.

All the practical difficulties which affect research in one prison are compounded when surveys are extended to other prisons. Negotiations must be conducted with a new set of personnel, and travel is added to the other costs. This explains why studies are often based on a single prison, even though it is unlikely to be representative of all prisoners. This problem applies particularly to female prisoners. Most British studies of women in prison are in fact studies of women in HMP Holloway, presumably because of its convenience for London-based researchers.

Theoretical and ideological problems

There is strong opposition to the principle of attaching psychiatric labels to prisoners. This opposition arose from the anti-psychiatry movement of the 1960s, which saw psychiatrists as agents of social control. These ideas were developed in the work of Foucault, Laing and Szasz, but the most articulate exponent of their application to prisons is Sim (1990). He portrays offenders as rebels, subversives who threaten the social order. Women who offend are a particular threat to the stable, nuclear family which is the cornerstone of society. According to Sim, psychiatry is an important weapon in the armamentarium of the prison, as it struggles to regulate, discipline and normalize female prisoners/rebels.

Sim does not claim to be impartial and Dell (1991) draws attention to the fact that he refers only to material that supports his thesis. Nevertheless, similar views are widely held, and many writers have criticized attempts to 'psychiatrize' offenders. It is suggested that one form of deviance (breaking the law) is transformed, by the intervention of doctors, into another form of deviance (the sick role). The latter implies a different form of response, treatment rather than punishment, which many would see as humane and positive. However, there is also a negative side. The sick role in psychiatry can also imply a negation of the individual's autonomy, and loss of rights that are retained by the law-breaker.

The concept of psychiatrization is a frequent theme in discussions of female offending and psychiatry, and it will be discussed in depth in later chapters. For the moment, its importance lies in directing our attention to the motivation of researchers who attach psychiatric labels to prisoners. The studies described in this chapter cannot be understood without asking the question: 'Why look at psychiatric disorder in prisoners?' The aims of studies vary, and there have been recent changes in emphasis.

Why study psychiatric disorder in prisoners?

Two answers to this question will be considered. The first is an attempt to discover the causes of crime, the second to evaluate psychiatric services.

Looking for the causes of crime

In reading many descriptions of psychiatric disorder in prisoners, it soon becomes apparent that their concern was less with prisoners than with criminals in general. The authors were attempting to explain offending, by describing its links with mental disorder. Prison was a convenient place in which to find and interview criminals. Little attention was paid to the type of prison, or the problems of obtaining representative samples and accurate prevalence rates.

Modern criminology has discredited this approach. Much crime is neither detected nor prosecuted, and prisoners are not representative of offenders in general. The best studies of offenders use samples drawn from the community (e.g. West and Farrington, 1973). As a result of studies of this type, it is now recognized that mental disorder accounts for only a tiny proportion of all offending. Attempts to link crime and personality characteristics in a general way are usually disappointing (although very important in certain individuals). In the search for general causes of crime, the focus of research has now shifted away from the mind of the criminal to social or situational determinants of offending. Criminality is more usefully seen as a variety of different behaviours occurring in specific situations, rather than an individual characteristic, of which some individuals have more and some less.

The net result of these developments is a recognition that prison studies can play only a minor role in attempts to describe the relationship between crime and psychiatric disorder. At the same time, the importance of prison studies has increased because they now have a different purpose. The emphasis has shifted away from a search for the causes of crime to an attempt to evaluate psychiatric services.

Evaluating psychiatric services

In any comprehensive account of psychiatric services, prison surveys are relevant in two ways. First, the extent of psychiatric disorder found in prisons is an important determinant of health care needs, when planning services for prisoners. Second, the number of seriously mentally disordered prisoners is a measure of the effectiveness of the National Health Service, in its dealings with mentally disordered offenders. One of the aims of a psychiatric service is to keep the seriously mentally disordered offender out of prison, so the number of mentally disordered prisoners gives an indication of how many are slipping through the net.

This change in the purpose of prison studies requires a change in methods. A prison survey designed to estimate the need for psychiatric services, or to measure their efficacy, will be different from one which was concerned only with the characteristics of criminals. The prevalence of disorders can only be measured using a representative sample, drawn at random from a defined population.

In obtaining a valid sample of prisoners, the first step is to distinguish between different types of prisoner and, in particular, to distinguish remandees (i.e. those who have not yet been tried, known as jail detainees in the USA) from sentenced prisoners. Suspected psychiatric disorder is one reason for remanding offenders in custody, even for trivial offences, so it is no surprise that previous studies of male prisoners have shown high rates of psychosis (Taylor and Gunn, 1984) and suicide (Dooley, 1990) in remandees. It is therefore impossible to make sense of prevalence rates without being clear about the population that has been surveyed.

Having defined the study population, criteria for identifying psychiatric disorder must be clearly stated, with cases described according to an accepted diagnostic system. In order to allow comparison of female and male prison populations, the same methods and criteria should be applied to both groups. Any departure from these methods means that the study is unlikely to yield useful prevalence rates, or allow a valid comparison of men and women. Previous studies of female prisoners will now be reviewed, with these strictures in mind. It soon becomes apparent that none of these studies allows a direct comparison of rates of psychiatric disorder in female and male prisoners.

Studies of female prisoners in England and Wales

This account is confined to studies in England and Wales. Findings in other countries, with different health care and criminal justice systems, are of limited relevance. However, a subsequent section makes brief reference to some North American surveys, as examples of good practice which allow valid comparison between women and men. A third section discusses specific problems in studies of mental handicap in prison.

Psychiatric studies of prisoners can be divided into two groups: those which describe the general prison population, and those which describe selected groups of women.

Studies of the female prison population

The psychiatry of the English female prison population is, in large part, the psychiatry of HMP Holloway: few published studies sample other women's prisons. An exception is the work of Epps, during her time as a medical officer at a Borstal in Aylesbury, Buckinghamshire. Epps and Parnell (1952) compared 177 of the Borstal women with 123 female undergraduates at Oxford, for both physique and temperament. This exercise is typical of early prison studies, using Borstal women as a convenient source of criminals for comparison with 'normal' women. The aim was to explain offending in terms of individual differences. The paper is of no more than historical interest. It includes an early attempt to apply similar measures to female and male prisoners but the scale chosen (Sheldon's 'psychiatric index') is of no value as a measure of psychiatric disorder.

A better description of psychiatric disorder emerges from Epps' (1951, 1954) survey of all girls (sic) sentenced to Borstal training between April 1948 and August 1950. Three hundred were included in the study, 275 being followed up over a prolonged period, and the remaining twenty-five seen once only for an interview and intelligence test. Although most of the paper is concerned with social characteristics and an attempt to describe the nature and causes of criminality, some psychiatric data are provided. Four women (1.3 per cent) were diagnosed as suffering from schizophrenia and were transferred to hospital. Neurotic symptoms were shown by sixty-three women (21 per cent). Substance abuse was not recorded as a diagnosis, but was said to be common in the family histories of the women.

Epps used Raven's matrices (Raven, 1958) to measure intelligence, placing 50 per cent of women in the 'average' range, with 4 per cent of 'superior' intelligence and 46 per cent described as 'subnormal'. The term subnormal is misleading in this context. It refers to women in the lowest two of Raven's five categories of intelligence. Twenty-five per cent of the general population also fall within this range, so the term cannot be equated with mental handicap or learning difficulties (which affect less than 5 per cent of the general population).

The follow-up study (Epps, 1954) reported on 100 of the original women who were seen in Holloway prison after being recalled for unsatisfactory behaviour or reoffending. 'Emotional instability' and neurotic symptoms were more common in this recidivist group, but there was no apparent difference in the proportion of those with low intelligence or a record of psychiatric treatment. Epps suggested that the recidivist population contained a number of women with personality and neurotic problems who required some form of psychiatric treatment, in addition to Borstal training, if they were to modify their offending behaviour.

Epps' study is a good illustration of research done by a serving prison medical officer, using prisoners as an example of offenders in general and attempting to understand offending in terms of psychological characteristics. It has many methodological flaws (none of the psychiatric terms are defined) but it makes important points. Psychosis is unusual, whereas personality disorder and neurotic symptoms are common, and women with the latter problems appear prone to reoffend. Epps assumed that the personality or neurotic problems were causes of recidivism but, even if the reverse were true, it is difficult to disagree with her assertion that these women required some form of help in addition to the ordinary custodial regime. Although her study began with a search for causes of crime, it ended by making sensible suggestions about services for women in prison.

In the same tradition, Woodside (1962), a social worker, described all women sentenced to six months or longer who were received at HMP Holloway during a six-month period in 1961, a total of 139 women. The study was dependent on previous reports and precise diagnoses were not given for the total of sixty-eight women (nearly half the sample) identified as 'unstable'. Fifty-eight (42 per cent) had a 'positive psychiatric history', including inpatient treatment in twenty-six (19 per cent). Five women had been 'certified patients' in the past and a total of twelve women (9 per cent) were 'subnormal'. The criteria for subnormality are vague, but seem closer to modern concepts of learning difficulties than those used by Epps.

Woodside made no direct comparison with male prisoners but the survey derived its inspiration and some of its methods from the work of Roper (1950), who surveyed male sentenced prisoners in Wakefield, where he was medical officer. He found that 18 per cent had a 'positive psychiatric history', a considerably lower figure than the 42 per cent reported for women. It is difficult to draw more precise conclusions as Woodside's study lacks detail, and psychiatric disorder is loosely defined.

The most influential survey of female prisoners was carried out by Gibbens (1971), who surveyed every fourth reception into Holloway during 1967, a total of 638 women. This work is regarded by some authors as the empirical basis for the psychiatric influences on the rebuilding of Holloway. Despite its pre-eminent position, this study was never fully written up, and existing accounts do not specify any criteria for the definition of psychiatric problems. The findings vary according to the type of prisoner (Table 2.1), but the author concludes that women in prison have high rates of psychiatric disorder.

Comparison of Gibbens' figures with those from a survey of sentenced men (Gunn et al., 1978) is not straightforward. A history of inpatient psychiatric treatment was found in 12 per cent of men compared to 17 per cent of women reporting such treatment within the last three years. On the other hand, 15 per cent of men had 'a drink problem', compared to 7 per cent of sentenced women described as suffering from alcoholism. While mental health was 'a major problem' in 15 per cent of sentenced women, 34 per cent of sentenced men were identified as current psychiatric cases.

These studies use very different methods and two particular problems limit the usefulness of direct comparisons between their findings.

Sampling

Gibbens used a reception sample whereas Gunn et al. studied a cross-section of men serving a sentence. The former method includes a high proportion of prisoners serving short sentences, the latter a higher proportion of long and medium term prisoners.

The definition of psychiatric disorder

The studies used quite different criteria for identifying psychiatric disorder. Some findings are comparable (e.g. reports of previous psychiatric treatment) whereas others, such as rates of alcoholism,

Table 2.1 Psychiatric disorder in women prisoners, by type of prisoner (adapted from Gibbens, 1971)

	Sentenced N = 128 %	Medical remands N = 57 %	Other remands N = 485 %
Mental illness	15	39	25
Psychiatric inpatient in last 3 years	17	26	18
Psychiatric inpatient > 3 years ago	8	10	6
Past suicide attempt	22	26	22
Alcoholism	7	7	3

Note: All figures are percentages of the total in that category of prisoner. Figures under 'other remands' were calculated as the mean of the figures given in the original table under 'Remand before conviction' and 'Remand after conviction'. The raw figures given as 'n' for each category were calculated from the percentages given in the original paper. In total, the percentages add up to 115 per cent, so the raw figures must be regarded as rough estimates.

cannot usefully be compared as they depend to a great extent on the criteria used to define the disorder.

These problems invalidate a direct comparison between the two studies. Unfortunately, no subsequent study has improved on Gibbens' work. A survey based on a random sample of 708 receptions to HMP Holloway (Turner and Tofler, 1986) found that 125 women (18 per cent) had a history of psychiatric treatment and 195 (28 per cent) had a history of self-harm. It is difficult to make sense of these findings because the authors give no diagnoses, apart from drug and alcohol abuse. Women with 'a history of psychiatric treatment' include those who received brief outpatient treatment for depression, alongside those who may have spent years in hospital suffering from schizophrenia. A further problem is the use of a reception sample, with no information about criminal characteristics, or the mix of sentenced and remand prisoners. The authors make no explicit comparison with male prisoners, but state that rates of psychiatric morbidity in women admitted to Holloway are high.

Morris (1987) has compared these results to those of Gunn et al. to argue that women and men in prison have similar rates of psychiatric disorder. It is more accurate to say that differences between the studies, in sampling and the definition of psychiatric disorder, preclude any meaningful comparison.

There have been no other general psychiatric surveys of women in prison since Gibbens' work. It is ironic that the heyday of surveys and research by prison medical officers appears to have passed, at the same time as there is increasing outside interest in mentally disordered prisoners. The early work was doomed to failure, as causes of crime could not emerge from studies of mental disorder in prisoners. By contrast, present concern centres on the operation and efficiency of psychiatric services. Questions of this type are ideally suited to research projects in individual prisons, carried out by doctors who have local knowledge of the service and its failings. It is time for prison doctors to resume their involvement in research, now that realistic questions are of clear and pressing importance.

Studies of selected groups of women in prison

Comprehensive surveys of female prisoners may be rare, but there are many published reports on selected groups of women in prison (usually in Holloway). The groups are chosen in the expectation that they will be of psychiatric interest, and they are not meant to be representative of female prisoners as a whole. There is no similar collection of studies of male prisoners, probably reflecting an assumption on the part of researchers that women in prison are more appropriate for study by psychiatrists.

The women described in these studies are selected in one of two ways. First, they may be chosen because they display a particular behavioural or psychiatric problem. Second, they may be chosen because of the nature of their offence, which may be uncommon or suggestive of mental disorder.

Women with particular behavioural or psychiatric problems
Early studies of drug use and crime made use of prison samples. D'Orban (1970, 1973) described sixty-six female drug users in Holloway, including a four-year follow-up. As the author's main concern was to describe the relationship between drug dependence and offending, prevalence rates were not measured. Most of the women were remand prisoners during the original study and some entered Holloway on more than one occasion. During the study period, the women faced a total of ninety-six charges. None were charged with the sale of drugs and twenty-nine cases (30 per cent) resulted in a custodial sentence. The sentence length was twelve months or over in only three cases, and never exceeded eighteen months.

Although service provision was of secondary interest, previous treatment is described. Seventeen per cent of the women had a history of inpatient psychiatric treatment preceding their addiction, and 50 per cent had received inpatient treatment since becoming addicted. Apart from drug addiction, most admissions were for personality disorder. Twenty-one of the ninety-six court disposals (22 per cent) resulted in hospital admission, six under the Mental Health Act category of psychopathic disorder. This pattern of disposal is noted by d'Orban to be similar to that for male addicts in prison.

The follow-up study found a strong association between continued addiction and continued delinquency, concluding that these findings were in accord with the view that addiction and crime are not causally related but are 'parallel effects of underlying factors leading to social deviance'. These papers provide a useful description of drug-dependent women passing through the prison system in the early 1970s, before the expansion in heroin use that occurred in the 1980s. In Chapter 8, these women will be compared to drug users identified in the present survey.

Cookson (1977) described self-injury by women in Holloway, and found that women who injured themselves tended to be younger than average, with longer sentences and more violent offences.

Incidents occurred in runs or epidemics, suggesting that imitation was important. The rate of self-injury was believed to be high, but was not compared to the rate in male prisons. The question of the relationship between self-injury and suicide was not addressed. This survey begs questions about the psychiatric diagnosis and treatment history of women who self-harm, but these issues were not addressed.

A more recent study by Wilkins and Coid (1991) found that 7.5 per cent of all remands to Holloway had a history of previous self-injury by cutting. Seventy-four women with a history of self-injury were described in more detail. Compared to remanded women with no history of self-mutilation, they were more likely to have had a disturbed and deprived childhood, and showed high rates of behavioural disturbance, substance abuse and personality disorder.

Women charged with a particular offence
Woodside (1961) described the twenty-six women admitted to Holloway during February 1960 on drink-related charges. They were found to consist mainly of homeless, alcoholic women with multiple social problems and extensive histories of petty offending.

The other papers in this category are a series written by d'Orban when he was working as a visiting psychiatrist at HMP Holloway. His papers are probably the best examples of this approach, containing both a detailed description of what he found in the course of his work, and an attempt to relate his findings to wider questions about the operation of psychiatric services and the criminal justice system. D'Orban (1972) described thirteen women remanded to Holloway for stealing a baby. They could be placed in one of four groups: women of low intelligence who stole a baby to play with (two cases); women suffering from schizophrenia (three cases); 'psychopathic personalities' with a history of delinquency and a preoccupation with the desire to have children (four cases) and a 'manipulative' group, where the offence was an attempt to influence the emotions of a third party (four cases). Four women (three from the manipulative group and one from the psychopathic group) were given a prison sentence for this highly unusual offence.

D'Orban (1979) described the eighty-nine women remanded to Holloway over a six-year period charged with attempted or actual filicide. Psychiatric disorder was found in seventy-five women (84 per cent), including fourteen women (16 per cent) suffering from psychosis. Rates of psychiatric disorder are high, but the selective nature of the sample precludes generalization to other women in prison. Only nine of the eighty-four women found guilty (11 per cent) received a prison sentence, and none of these were in the 'mentally ill' group. It is not possible to determine the precise psychiatric characteristics of the women who did receive a prison sentence, but the results are consistent with a reasonably effective mechanism for screening out mentally ill women for a disposal other than imprisonment. The question remains of how effective such mechanism may be in detecting and diverting those women charged with more mundane offences.

A third paper by d'Orban (1985), described the seventy-two women sentenced to imprisonment for contempt of court and received at Holloway prison from 1979 to 1983 (45 per cent of all women in England and Wales imprisoned for contempt during this period). Psychiatric disorder, according to clinical criteria, was found in twenty-seven women (38 per cent). The diagnoses were schizophrenia (eight cases), paranoid state (four), personality disorder (nine), reactive depression (two) and alcoholism (four). As there are no studies of psychiatric aspects of contempt of court in men, it is unclear to what extent the high rates of psychosis are a feature of this uncommon offence in both sexes. While the study shows high rates of psychiatric disorder in a particular group of sentenced prisoners, they were civil prisoners (who accounted for only 0.1 per cent of the sentenced female prison population in 1988) and legislation (since amended) relating to the particular offence of contempt did not allow for a remand for psychiatric reports.

These reports make no claim to be representative of female prisoners in general, but it is possible that their presence in the literature contributes to the impression that female offenders have a high level of psychiatric disturbance. Some of the studies appear to have been inspired by the assumption that women in prison are likely to have high levels of disorder and there is a lack of corresponding studies on groups of male prisoners.

Studies in the United States

American studies are of limited relevance because of the different ways in which their criminal justice and prison systems are organized. They remain of interest because they demonstrate two features which are absent from studies in the United Kingdom:

1. Application of the same method to samples of both female and male prisoners.
2. An attempt to relate prevalence rates of psychiatric disorder in prisoners to rates in the community.

Guze (1976) compared 223 male and 66 female sentenced prisoners, using a standardized interview. The male sample consisted of all inmates in Missouri becoming eligible for release during the study period (November 1959 to April 1960). The female sample consisted of all women under parole supervision on 1 July 1969, plus seven women released during the course of the study. The ten-year gap between collection of the samples means that the male and female samples are not directly comparable, but the results are presented in Table 2.2.

The authors conclude that 'the prevalence of sociopathy, alcoholism and drug dependence was similar to that among male felons'. This statement is correct if these diagnoses are combined, but masks the significantly greater prevalence of drug dependence in female prisoners (the odds ratio is 7.39, with a 95 per cent confidence interval of 3.2 to 17.1). Hysteria (the definition is of somatization disorder or Briquet's syndrome) was frequent in female prisoners but was not found in any males. The reported differences in the prevalence of 'mental deficiency' are discussed below.

Apart from the recording of homosexuality as a psychiatric disorder, other peculiarities of the study deserve comment. At least one psychiatric diagnosis was given to all the women and 90 per cent of the men. These figures are much higher than those reported in British studies, and reflect the use of different diagnostic criteria. The American 'Feighner criteria' use operational definitions for high reliability, but can be criticized for not reflecting the reality of

Table 2.2 Psychiatric disorder in prisoners, by gender (from Guze, 1976)

	Women N = 66 %	Men N = 223 %
Sociopathy	65	78
Alcoholism	47	54
Hysteria	41	–
Drug dependence	26	5
Homosexuality	14	1
Anxiety neurosis	11	12
Depression	1	–
Mental deficiency	6	< 1
Schizophrenia	1.5	1
Epilepsy/organic disorder	–	2
Uncertain	12	< 1
Any diagnosis	100	90

psychiatric practice. There are considerable doubts about the face validity of these findings. Sociopathy is the most common diagnosis in both sexes and it is unlikely that the 65 per cent of women and 78 per cent of men with this diagnosis would be regarded as true psychiatric cases. It is certain that many would not be receiving psychiatric treatment and the relevance of the diagnosis to service provision is unclear.

Another surprising finding is the low frequency of depression. This was noted only when it was a 'primary diagnosis' but one case in a total of 289 subjects remains a low prevalence for a disorder which is common in most populations. The very high prevalence of hysteria is also unexpected.

Novick, Penna and Schwarz (1977) described the 1300 men and 120 women who comprised all receptions to New York City Correctional Facilities during a two-week period in 1975. Diagnoses were derived from the non-standardized, routine interview administered to all receptions by primary care physicians.

A history of drug use (mainly use of heroin or methadone) was given by 48 per cent of women and 40 per cent of men. Eight per cent of women and 9 per cent of men were judged by the admitting physician to be in need of psychiatric evaluation. 'Psychiatric disorder' was recorded as a 'diagnosis or problem listing' in 10 per cent of women (compared to 5 per cent of males under 21 years and 7 per cent of males 21 years and over). Corresponding figures for alcohol abuse were 4 per cent of women (0.3 per cent and 7 per cent of younger and older men respectively) and for drug abuse 23 per cent of women (13 per cent and 17 per cent of men). Five per cent of women and 7 per cent of men reported previous psychiatric hospitalization. A history of attempted suicide was given by 8 per cent of women and 5 per cent of men. Following the medical examination, 5 per cent of the women and 7 per cent of the men were sent to 'mental observation areas'.

Sampling in this study was comprehensive and permits comparison of female and male inmates, but the absence of precise diagnoses makes it difficult to draw any useful conclusions. The higher prevalence of drug abuse and dependence in women is consistent with the previous study.

A later study of female prisoners in Missouri (Daniel et al., 1988) examined 100 consecutive sentenced receptions, using the Diagnostic Interview Schedule (Robins et al., 1979) to generate DSM–III diagnoses (American Psychiatric Association, 1980). Ninety women received at least one diagnosis on Axis I, with sixty-seven receiving more than one. This study did not examine male prisoners but the authors compare their findings to rates of psychiatric

Table 2.3 Lifetime prevalence of psychiatric disorder in women in prison and in the general population of St. Louis (from Daniel et al., 1988)

	Population sample N = 1802 %	Prison sample N = 100 %	p*
Schizophrenia	1.1	7	0.0003
Major depression	8	19	0.001
Mania	1.1	2	ns
Alcohol problems	4	36	0.001
Drug problems	4	26	0.001
Simple phobia	9	15	ns
Agoraphobia	6	6	ns
Panic disorder	2	2	ns
Obsessive compulsive disorder	3	6	ns
Somatization	0.3	1	ns
Antisocial personality disorder	1.2	29	0.0001

* p derived from Fisher's exact test or Chi^2, as appropriate.

disorder found in the community. Table 2.3 shows the lifetime prevalence of psychiatric disorder in female prisoners compared to rates reported for females in the general population of St. Louis.

The authors note that for every comparison where the differences are significant, prisoners have higher rates of psychiatric disorder. They go on to demonstrate that these significant differences remain when age-specific prevalence rates are used, i.e. they cannot be explained away as the result of the different age structure of the two populations.

Attention is drawn to the high rate of schizophrenia and major depression compared to most other studies. The high prevalence of depression may be due to the use of a reception sample, as prisoners react to their recently imposed sentence and adjust to losing their liberty. These factors cannot account for the high rate of schizophrenia but the prevalence rate of 7 per cent derives from a small sample, where it corresponds to seven cases. The 95 per cent confidence interval is from 3 per cent to 14 per cent, i.e. the findings are compatible with the possibility that the true rate of schizophrenia in this section of the Missouri prison population is 3 per cent. The interpretation of this finding must also depend on a knowledge of local practice in respect of mentally disordered offenders.

Perhaps the most important point made by this study is that women in prison differ in many ways from women in the general population, including having higher rates of most psychiatric diagnoses.

Studies of mental handicap in prisoners

Although it has become much less fashionable in recent years, it is probably still true that 'more has been written about the association of low intelligence and criminal behaviour than any other factor' (Bluglass, 1966). It is therefore surprising that the extent of mental handicap among prisoners remains unclear, and the literature is very confusing.

One reason for the confusion is that, in the early part of this century, it was widely believed that mental retardation and criminality were directly related, presumably reflecting an assumption that criminality was inherently irrational. Lewis (1974) emphasized the close relationship between the historical development of the concept of psychopathic disorder and low intelligence. This theoretical confusion has been compounded in epidemiological studies by a failure to distinguish between the concepts of low IQ on the one hand, and mental handicap on the other. As a result, a recent review of studies of psychiatric disorder in sentenced prisoners (Coid, 1984) was able to quote prevalence rates for mental handicap ranging from less than 1 per cent to 45 per cent. For this reason, mental handicap requires separate discussion, although many of the studies have already been mentioned above. Two specific issues arise:

1. The definition of mental handicap and ways of estimating its prevalence.
2. The relationship between mental handicap and other psychiatric disorder.

The definition and prevalence of mental handicap

It is necessary to distinguish low intelligence, essentially nothing more than a test score, from the diagnosis of mental handicap.

The definition of low intelligence
This is relatively straightforward. If intelligence is defined by a score on an IQ (Intelligence Quotient) test, then low intelligence is simply a score falling below an arbitrary point. Raven's matrices (Raven, 1958) are an alternative measure of intelligence, often used in prison surveys.

Mental handicap or mental retardation
The definition of mental handicap is more complex. The approach taken in the International Classification of Impairments, Disabilities and Handicaps (Wood, 1980) is preferred by specialists in

the field (Campbell, 1990), and the definitions used illustrate potential sources of confusion:

An **impairment** is an abnormality or loss of any structure or function.

A **disability** is any restriction or lack of ability to perform an activity in the manner or within the range considered normal for a human being, resulting from an impairment.

A **handicap** is a disadvantage for a given individual, resulting from a disability, that limits or prevents the fulfilment of a role that is normal for that individual.

The definition of a handicap contains a social component. Failure to score within the normal range on an IQ test is a disability but it is not a good predictor of other disabilities, and it is a very poor predictor of handicaps, i.e. problems of social functioning (Hunter, 1979). The social component in handicap is recognized in the World Health Organization's (1979) definition of mental retardation, which stresses that an assessment should include information about 'adaptive behaviour': IQ levels are provided as a guide, with the proviso that they 'should not be applied rigidly'.

These distinctions suggest three possible epidemiological measures:

1. The prevalence of low intelligence, as measured by IQ tests.
2. The prevalence of mental handicap, assessed by standardized measures which take account of social functioning.
3. The prevalence of mental handicap, as defined by presentation to mental handicap services.

This third measure is closely related to the concept of administrative prevalence, defined by Tizard (1964) as 'the number for whom services would be required in a community which made provision for all who needed them'. In the general population, the prevalence of low intelligence (defined as IQ less than 70) is 20–30/1000 population, whereas the administrative prevalence of mental handicap is less than 10/1000, as less than half of all people with low intelligence require services. For the pur-pose of estimating the need for services, administrative prevalence is most useful.

Mental handicap and psychiatric disorder

Populations of mentally handicapped people have increased psychiatric morbidity. Rates of psychosis, neurosis and personality disorder are elevated but diagnosis is often difficult, as the manifestations of psychiatric disorder in a mentally handicapped person may fit poorly into diagnostic categories developed in a population of normal intelligence (Corbett, 1979). There may be particular difficulties in differentiating disturbed behaviour resulting from situational factors, and that caused by psychiatric disorder. Such decisions may require a prolonged multi-disciplinary assessment, which is beyond the scope of most prison surveys. Therefore, there is a greater risk of diagnostic mistakes in prisoners with a significant mental handicap.

Previous surveys of mental handicap in prisoners

Most existing surveys are of male prisoners, but they serve to illustrate the general problems in this area. The association between delinquency, low IQ and poor educational attainment is a robust finding (Farrington, 1990) and there is good evidence that the association between low IQ and offending remains significant, though less strong, after controlling for social factors (Hirschi and Hindelang, 1977). Therefore an increased prevalence of low intelligence in the prison population is to be expected, whereas a high prevalence of mental handicap would be a cause for concern, suggesting that existing screening and diversion procedures were ineffective.

Many surveys of prison populations have measured the prevalence of low intelligence, while using a terminology derived from mental handicap. Roper (1950) found that 50 per cent of inmates at Wakefield prison were 'below average/dull' and 18 per cent were described as 'intellectually defective/very dull', according to the categories in Raven's matrices. These results are shown alongside those for the general population in Table 2.4.

While the figures for the prison population are skewed towards the lower end of the range, it is obvious that the 'below average' categories cannot be considered synonymous with mental handicap. Roper did not consider social functioning in relation to the test results although, elsewhere in the paper, he states that half a per cent of his sample were 'profoundly intellectually defective' and were therefore excluded from his analysis.

Epps (1951) tested 289 women using Raven's matrices and describes 144 (50 per cent) as falling in the 'average' category with 134 (46 per cent) of 'subnormal' and 11 (40 per cent) of 'superior' intelligence. Assuming that these figures result from merging the two lowest and two highest categories respectively, they resemble Roper's findings and justify Epps' assertion that 'the intelligence

Table 2.4 Scores on Raven's matrices of sentenced female prisoners (from Woodside, 1962), sentenced male prisoners (from Roper, 1950 and Bluglass, 1966) and the general population

	Female prisoners (Woodside) N = 118 %	Male prisoners (Roper) N = 814 %	Male prisoners (Bluglass) N = 291 %	General population %
Very superior	11	1	7	5
Superior	16	4	24	20
Average	31	45	47	50
Below average	24	32	16	20
Intellectually defective	18	18	6	5

of the average delinquent is less than that of the average non-delinquent'. The relevance to mental handicap is uncertain.

Woodside (1962) tested 118 sentenced women on Raven's matrices (also shown in Table 2.4) and concluded that they showed a 'slight loading towards the lower intelligence grades'. In her total sample of 139 women, she identified twelve (9 per cent) as subnormal. The author accepted that her diagnostic criteria were 'rough and ready', but noted that five of these women had been psychiatric inpatients, another three had been 'certified patients in institutions' and a further two had been recommended for a hospital order under the Mental Health Act, 1959. She appears to have found a relatively high percentage of 'true' mental handicap in a prison population.

Bluglass (1966) combined Raven's matrices, the Mill Hill vocabulary test (Raven, 1943) and a clinical assessment in his survey of 300 male inmates in Perth prison. The scores on Raven's matrices are included in Table 2.4 and approximate to the distribution for the general population. His criteria for mental handicap resemble the second type of prevalence measure described above, i.e. a standardized measure which also takes account of social function. He placed 11.6 per cent of the sample in the borderline range and identified 2.6 per cent as subnormal.

Guze's (1976) survey of 223 male prisoners identified 7 per cent as 'below normal intelligence' on the basis of school history and mental state, plus test results when these were available from records. IQ scores were available for three of these subjects and fell between 80 and 85; only one subject was thought to have a degree of impairment that would preclude a normal, self-sufficient life in the community. Of the sixty-six women in this study, four (6 per cent) were considered 'mildly retarded' though 'none seriously enough to interfere with daily life'. Only one woman had test results, an IQ of 60.

Chapter summary

Studies of prisoners, both female and male, tend to show a higher rate of psychiatric disorder than would be expected in the general population, but psychosis accounts for only a small proportion of total morbidity. A number of studies suggest that women in prison in England and Wales have higher rates of psychiatric disorder than men in the same position. However, there is inconsistency between studies, and the definitive survey, applying identical methods to equivalent samples of women and men, has not been done. The literature on mental handicap or learning difficulties is confusing but most studies suggest that, while the distribution of intelligence in prisoners is skewed towards the lower end of the scale, only a small proportion would meet diagnostic criteria for mental handicap. Only one study, by Woodside, suggests that the prevalence of mental handicap in female prisoners may be as high as 9 per cent.

The following chapters describe a survey of the 1200 women who make up the entire female sentenced population of England and Wales. The authors surveyed the male sentenced population, using similar methods, to allow a valid comparison of rates of psychiatric disorder.

3

A psychiatric profile of the female prison population

Introduction

This chapter describes the design and methods of a survey of women serving a prison sentence in England and Wales. It formed part of a larger study of psychiatric disorder in sentenced prisoners, undertaken by the Department of Forensic Psychiatry at the Institute of Psychiatry, between 1988 and 1990 (Gunn, Maden and Swinton, 1991). The research workers and interviewers were the author and Dr Mark Swinton, both working under the supervision of Professor John Gunn. The project was funded by a grant from the Directorate of the Prison Medical Service via the Home Office's Research and Planning Unit.

The first aim was to describe the prevalence of psychiatric disorders within the sentenced prison population, in both men and women. The second aim was to estimate the treatment needs of prisoners suffering from psychiatric disorder. In order to achieve these objectives, the basic design was an interview and case note study of a cross-sectional sample comprising one quarter of all women serving a prison sentence in England and Wales, and about 5 per cent of men in the same position.

It is inevitable that much of this chapter is of a technical nature, being intended to establish the scientific credentials of the survey. The general reader may wish to skip some of these sections.

The prison population of England and Wales

At the time of the survey, in 1988 and 1989, the prisons of England and Wales held 1776 women, including 515 who were remanded awaiting trial or sentence and were therefore excluded from the study. Of the remaining 1256 women, twenty-seven were serving a prison sentence only in default of payment of a fine and were also excluded. This left 1229 women, sentenced by the courts to immediate imprisonment, as the population to be surveyed.

In contrast, the male population on the same date contained 37 292 adults and young offenders serving a sentence. After the exclusion of 549 fine defaulters, 36 743 men were left in the population to which female prisoners would be compared. The ratio of women to men is 1:30. It is important to remember that this survey was confined to sentenced prisoners, with remanded prisoners excluded from both the male and female samples. As a result, both samples are biased towards more serious offenders. A sentence of imprisonment is the most severe sentence that our courts can impose and many offenders, even when prosecuted, never reach the sentenced prison population.

The study sample

The prisons

Eight prisons in England and Wales hold sentenced women and the study was carried out in four of them: Holloway, Styal, Drake Hall and Durham.

Holloway Prison
A closed remand and allocation centre serving the south-east of England, Holloway has an average daily sentenced population of around 120 women. Some are in transit, awaiting allocation to another prison, while others will serve most or all of their sentence here for various reasons:

1. Sentenced women provide the inmate workforce of the prison.
2. The prison has good 'hospital' facilities and holds sentenced women who need a high level of psychiatric care, particularly if efforts are being made to arrange transfer to an outside hospital.
3. It has a mother and baby unit.

The original sample included sixty-four women in Holloway.

Styal Prison
A closed training prison in the north-west of England which holds a general cross-section of women serving their sentences. It has full-time medical cover and regular visiting psychiatrists. Specialist units include a mother and baby unit, a 'hospital' and a vulnerable prisoner unit.
The original sample included 120 women in Styal.

Drake Hall

An open training prison, representing the lowest level of security within the prison system. The 'hospital' does not have a full-time medical officer but is staffed by nurses who provide a ready source of help and advice. Inmates would not be allocated to open conditions if known to be mentally disordered or to represent an escape risk. It is relatively common for inmates to abscond from open prisons, resulting in their return to a closed institution.

The original sample included ninety-six women in Drake Hall.

Durham Prison H Wing

A closed, high security wing for long-sentenced prisoners, situated within the perimeter of a male prison. It holds approximately forty women and all female life prisoners spend the first part of their sentence here, usually around three years. Women who require a high level of security remain in Durham for longer periods as it is the only high security facility for women. At the time of my visit, the longest serving woman had been there for fifteen years.

The small size and stable population present a number of problems. All work is provided on the wing and is limited in scope. Both staff and inmates commented on the claustrophobic atmosphere.

The 'hospital' consisted of the top landing of the wing. There is a full-time medical officer and regular sessions by a visiting psychiatrist. The nurses who are always present in the 'hospital' were seen by the inmates as helpful and sympathetic and, to some extent, independent of the disciplinary structure of the prison.

The original sample included twenty-one women in H Wing.

Sample characteristics and sample : population matching

The four prisons visited hold about half of all women serving a prison sentence. Sampling was carried out within each prison by selecting every second woman from an alphabetical list. The end result was a one in four (or 25 per cent) sample of the female sentenced population, that spanned the full range of security levels and had a good geographical spread.

The sampling procedure was designed to avoid bias, and to produce a representative group of women. In order to confirm that the sample was representative, a number of comparisons were made between the sample and the characteristics of the total sentenced population, as reported in official prison statistics. These tests are described in full, for both the female and male samples, in Gunn,

Maden and Swinton (1991). Only three will be described here: offence type, sentence length and ethnic origin.

Offence type

Table 3.1 shows the offences for which women had been sentenced, for the sample and the prison population as a whole.

The percentages are fairly similar. One of the main differences is in the 'other' and 'unrecorded' categories, where the researchers were able to be more precise than the official statistics (the offence is always recorded in the prison record of each inmate). The figures confirm that most women (and men) are sentenced for theft or other acquisitive offending. However, drugs offences account for almost one-third of all women serving a sentence, and violent offending is also important, accounting for one-fifth. In Chapter 1, it was stated that violence accounts for only 10 per cent of all indictable crime in women. This difference emphasizes the fact that sentenced prisoners are not representative of all offenders, as only more serious offences result in a prison sentence.

Sentence length

Table 3.2 shows the distribution of sentence lengths in the sample and total population.

The sample : population matching is not perfect, with an apparent over-representation of women serving long sentences, and an under-representation of medium sentences. Life sentenced prisoners are also over-represented in the sample, but the difference between the predicted 6 per cent and actual 8 per cent is only five individuals. Perfect matching would not be expected. Overall, it can be argued that the degree of matching is good enough to allow us to

Table 3.1 Female sentenced prisoners: comparison of sample and total population, by offence type

Offence	*Sample*		*Population*	
	n	%	n	%
Violence	54	18	247	20
Robbery	24	8	73	6
Burglary	18	6	57	5
Theft	92	31	313	26
Drugs	90	30	314	26
Other	23	8	185	15
Unrecorded	–	–	40	3
Total	301	100	1229	100

Table 3.2 Female sentenced prisoners: comparison of sample and total population, by sentence length

	Sample		*Population*	
	n	%	n	%
Short < 18 months	128	42	512	41
Medium 18–36 months	59	20	336	27
Long > 36 months	91	30	332	26
Life	23	8	76	6
Total	301	100	1256	100

assume that the sample is representative of all female sentenced prisoners in England and Wales.

Ethnic origin
The annual prison statistics contain very limited demographic information, but one of the most striking gender differences is in the number of prisoners from ethnic minorities. The 1988 figures showed that 25 per cent of sentenced women were from ethnic minorities, with women of West Indian/African origin accounting for 20 per cent of the female prison population, compared to 9 per cent of males. The proportion of women from ethnic minorities also varies greatly by offence group. Five per cent of women serving a sentence for burglary are of West Indian/African origin, while they make up 40 per cent of women sentenced for drug offences. By way of qualification, it is pointed out in the Prison Statistics that these figures have 'limited explanatory value . . . in providing conclusive evidence, both as regards the involvement of particular ethnic groups in crime and in relation to the practice of the courts' (Home Office, 1988: 14). However, the statistical over-representation of women from ethnic minorities in certain parts of the prison population remains unexplained, and has been ignored in most writing on women and the criminal justice system (Cope, 1989). The published statistics suggest that the female sentenced population differs from the male sentenced population in containing an increased number of black women who have committed drug offences.

In the course of the interviews, it became apparent that many black women serving a prison sentence were drug couriers, usually from West Africa. They had no address in this country and were due for deportation at the end of their sentence. A description of drug couriers by Green (1991) covers their background, motivation and problems within prison. The Prison Statistics do not allow the identification of foreign nationals, but the present study included information about usual country of residence, and Table 3.3 relates this information to ethnic origin.

The table shows that the ethnic composition of the original sample of women is representative of the sentenced population as a whole. In total, thirty-nine women (13 per cent of the sample) were ordinarily resident overseas, including twenty-seven black and five Asian women who were serving sentences for the importation of drugs (usually heroin) and were to be deported at the end of their sentences. Considering only the 262 women ordinarily resident in the UK, the proportion of black women falls to 13 per cent and that of Asian women to 2 per cent.

Of the 1769 men interviewed, only eighteen (1 per cent) were ordinarily resident outside the UK. The ethnic origin of the remaining 1751 men was: white 1485 (85 per cent), black 186 (10.5 per cent), Asian sixty-two (3.5 per cent) and 'other' eighteen (1 per cent).

The criminological characteristics of women ordinarily resident in the UK will now be described in more detail. Asian women are included in the 'other' category as their numbers are so small. In those women for whom the information was available, Table 3.4 shows that 67 per cent of white and 61 per cent of black women were serving their first custodial sentence. This difference is not statistically significant. Eighteen per cent of both groups had served two or more previous sentences.

Having excluded overseas residents, there is now no particular concentration of black women among those convicted of drug

Table 3.3 Sentenced female prisoners: ethnic origin by country of residence in the study sample, with figures from the Prison Statistics as a reference

| | Overseas residents | | UK residents | | Total | | Prison statistics |
	n	%	n	%	n	%	%
White	5	13	219	84	224	74	72
Black	27	69	33	13	60	20	20
Asian	5	13	6	2	11	4	2
Other	2	5	4	1	6	2	6
Total	39	100	262	100	301	100	100

Table 3.4 UK resident female prisoners: number of previous custodial sentences by ethnic origin

	None	One	Two	Three or more
White	144	33	25	14
Black	20	7	5	1
Other	10	0	0	0
Total	174	40	30	15

offences (Table 3.5). The seven black women sentenced as drug offenders had been convicted of offences involving cannabis only. Ten of the white drug offenders had been sentenced for offences involving cannabis, but 33 cases involved opiates or amphetamines.

In contrast to these figures, none of the overseas resident women had any previous convictions. All but one (who had been convicted of spying) were serving a sentence for drug smuggling. None of these women had a history of criminality in their families, and all were employed in 'middle-class' occupations. Typically, they were businesswomen or traders who travelled regularly between their own country and London.

These findings have implications for the study. Women from other countries account for most of the over-representation of black women in the sentenced female population, compared to the male prison population. These women differ from UK residents in a number of ways, including the specialized nature of their offending, their lack of any criminal history, and the nature of their lifestyles. It is reasonable to assume that the pattern of psychiatric disorder in these women may be different from that of UK resident women. In addition, their future treatment would not be the responsibility of the health service. Only 1 per cent of male prisoners were resident overseas. Therefore a more valid and useful comparison of rates of

Table 3.5 UK resident female prisoners: index offence by ethnic origin

	White		Black		Other		Total	
	n	%	n	%	n	%	n	%
Theft/robbery	112	84	18	13	4	3	134	51
Drugs	43	83	7	13	2	4	52	20
Violence	43	80	7	13	4	7	54	21
Other	21	95	1	5	0	–	22	8
Total	219	84	33	13	10	4	262	100

psychiatric disorder may be obtained by excluding overseas residents from both samples. In Chapter 5, rates of psychiatric disorder in women resident overseas will be described separately.

Refusal rates

Each woman selected was approached by an interviewer. It was explained that the survey was being conducted by doctors from outside the prison system and would inquire about 'medical, psychiatric, drink and drug problems and the need for treatment'. An assurance was given that the interview was to be conducted in confidence, none of the information was to be made available to the prison authorities and the study would have no effect on the way in which she was managed within the prison system. Women who agreed to take part were required to sign a consent form.

Of 301 women approached, only eight (2.7 per cent) refused, leaving 293 subjects. Three of the eight 'refusers' were women from overseas and language difficulties made it impossible to explain the nature of the project, so the true or informed refusal rate was even lower and compares favourably with other psychiatric surveys, whether in prison or in the community. Several factors probably contributed to the low refusal rate. Prisoners are a captive population and any contact with outsiders provides a break from routine. Both interviewers were male and therefore had curiosity value in a predominantly female environment. Prisoners are rarely asked to comment on any aspect of their confinement and many women saw the interview as an opportunity to describe to sympathetic outsiders the care they had received.

Sampling of sentenced male adults and young offenders and matching with the male sentenced population are described elsewhere. A total of 1884 male prisoners were approached and 1769 agreed to be interviewed, a refusal rate of 6 per cent.

Data collection and interview

The object of data collection was to provide a demographic, criminological and behavioural description of each woman interviewed and then to identify and describe in more detail those who warranted a psychiatric diagnosis. Data were collected from five main sources:

1. Prison discipline records and prison medical records.
2. A semi-structured interview, including a standardized assessment of mental state using the Clinical Interview Schedule (Goldberg et al., 1970).

3. The woman's criminal record sheet, obtained from the Criminal Record Office.
4. Informal interviews with prison staff familiar with the interviewee.
5. Notes or reports from NHS hospitals when a history of previous treatment was obtained.

The first three sources of information were used for all women, adding the interviews with prison staff and notes from outside hospitals only where indicated. Information from these sources was used to complete a data sheet which had been designed for the study (see Maden, 1992 for a copy of the data sheet).

Each source of information will now be described in more detail.

Prison records
In prison, every woman has a discipline record and a medical record. Figure 3.1 lists the information which is always contained within the discipline record, while Figure 3.2 lists the additional information which appears in some inmates' records.

The inmate medical record (IMR) is a separate document which contains the initial screening and medical examination carried out on every inmate plus details of any contact with the doctor since then, including a record of past and current treatment. When mental disorder is suspected, the record may contain an extensive psychiatric history, copies of psychiatric reports prepared for the trial or relating to previous treatment and correspondence with psychiatrists outside the prison concerning possible transfer to hospital.

The interviewer was able to see these records and record the data before attempting to interview the woman concerned. As a result,

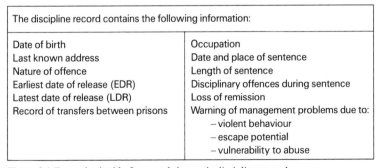

The discipline record contains the following information:	
Date of birth	Occupation
Last known address	Date and place of sentence
Nature of offence	Length of sentence
Earliest date of release (EDR)	Disciplinary offences during sentence
Latest date of release (LDR)	Loss of remission
Record of transfers between prisons	Warning of management problems due to:
	– violent behaviour
	– escape potential
	– vulnerability to abuse

Figure 3.1 Data obtainable from each inmate's discipline record

In addition, the discipline record *may* contain the following information:	
A detailed account of the index offence Social enquiry reports Psychiatric reports Full criminal record	Parole reports by: – prison officers – the governor – probation officer – education officer – chaplain

Figure 3.2 Data which may be obtainable from each inmate's discipline record

the interviewer had a considerable amount of background information at the start of the interview. The aim was to mimic clinical
practice in a hospital setting, where the doctor has at least a referral
letter before starting the assessment interview. This approach has
an important additional benefit in prison studies, by providing a
partial check on reliability. One paradox of such studies is that they
rely on self-report in individuals who may have a long record
of dishonesty. The prior availability of background information
allows the interviewer to draw attention to any inconsistencies
during the interview. In fact, it was rare for major discrepancies to
emerge, which must be seen as a tribute to the personal honesty of
the women concerned.

The interview
All interviews were carried out by a psychiatrist (the author or
Dr Mark Swinton), in private within the prison. A semi-structured
interview was designed for the project, including questions about
past and present medical and psychiatric problems, substance abuse
and self-harm. There was also a standardized assessment of mental
state, the Clinical Interview Schedule (CIS) (Goldberg et al., 1970).
This interview is designed for use in identifying psychiatric disorder
in general practice and consists of a number of stem questions
leading on to more detailed questions when the response is positive.
The questions concern mainly neurotic symptoms, e.g. 'do you ever
find that you get anxious or frightened for no good reason?', 'do
you worry a lot about things?' The CIS was also used as a convenient starting point for enquiry about other symptoms, including
abnormal experiences and beliefs, when appropriate.

The Schonnell reading test (Schonnell, 1961) was included, at the
interviewer's discretion, whenever there was a question of significant learning difficulties. This test gives a rapid estimate of reading
age. While a low score may be due to poor education, a high or

reasonable score is strong evidence against significant learning difficulties or mental handicap.

If there was any question of mental disorder or substance abuse being present, the interview was completed by a discussion of treatment possibilities. The aim was to establish the woman's attitude towards treatment, which was rated on a five-point scale used in a previous prison survey (Gunn et al., 1978: 265–7). The rating was arrived at by a process of negotiation between psychiatrist and interviewee, replicating a similar situation in the psychiatric clinic.

The interview could be completed in 30 minutes for subjects without evidence of psychiatric problems, and was expanded as necessary for others. Trained psychiatrists make expensive research interviewers but provide the important advantage of flexibility. A detailed history and mental state examination could be carried out when necessary, while minimizing the risk of offending women who gave no history of psychiatric disorder.

Interviews with prison staff
In specific cases, with the consent of the woman concerned, interviews with staff provided supplementary information to aid diagnosis. People in regular daily contact with the inmate were able to give useful descriptions of behaviour and relations with others. One difficulty with making psychiatric assessments in prison is the lack of a social context, and observations of behaviour outside the interview setting. In hospital psychiatry, the family would be used as informants to supply this information. An attempt was made to use prison staff in a similar way. The possibility of bias is obvious. However, it is probably no greater a danger than when family members are used as informants. Prison staff and inmates spend long periods in close proximity and relationships can become intense and complex, particularly in prisons which operate a personal officer scheme. These schemes provide each inmate with a named officer to whom the inmate should address both practical difficulties and more personal problems. In some cases, the role becomes that of a counsellor. Interviews with staff can therefore provide crucial information about how a woman is coping with prison and the use of informants in this way, with consent, followed the principles of good clinical practice.

Reports from outside hospitals
Whenever information was obtained about previous psychiatric contact, the woman concerned was asked for permission to contact the relevant hospital. This was never refused, and attempts were

made to obtain copies of notes or reports. Some women had notes documenting many years of psychiatric contact, or full assessments by their regional forensic psychiatry service.

The criminal record

A record of each woman's past convictions was obtained from the Criminal Records Office within the Home Office. This information was then summarized on each woman's data sheet.

Having assembled a mass of information from these five different sources, the final task is to coordinate the data in order to arrive at the whole point of the study: a psychiatric diagnosis and a decision as to the most appropriate treatment. The coding of data is described in the following section.

Rating and coding of data

For the background information, most of the coding was straightforward and was recorded on the data sheet during the interview. Other ratings require assessments and opinions and they were coded later, after obtaining further information when necessary. A coding manual (copies available from the author) provided guidelines for making judgments, including operational definitions whenever possible.

As a major aim of the survey was to estimate need for services, judgments about the significance and meaning of behaviour had to be made. Nevertheless, an attempt was made to record as much raw data as possible. For example, drug dependency was defined according to a set of operational criteria. In addition, all reported drug use was recorded, noting both frequency and type of drug use, without any attempt to assess the significance of this behaviour. Both sets of data could then be analysed, for different purposes.

The most important ratings concerned psychiatric diagnosis and treatment recommendations. The process by which we arrived at these ratings will now be described in more detail. The aim of each assessment was twofold. First, to assign a diagnosis according to an accepted classification system and, second, to make a decision on the most appropriate management.

Making a psychiatric diagnosis

The diagnostic system used was the Mental Disorders Glossary of ICD9, part of the World Health Organization's International

Classification of Diseases, Ninth Revision (World Health Organization, 1978). It is the most widely used psychiatric diagnostic system in Europe. It consists of a booklet providing brief descriptions of the features of each disorder, and a four-digit code number. The definitions are widely accepted and both interviewers had previously used the system in their clinical practice. The diagnosis was based on clinical criteria, with reference to 'present state' rather than lifetime diagnoses, i.e. the critical decision was whether or not a woman was suffering from a psychiatric disorder now, rather than whether she had ever suffered from psychiatric disorder in the past. The exception was substance abuse. In principle, at least, alcohol and drugs are not available within prison, so the reference period was the six months prior to arrest, when last at liberty.

Neither the interview designed for the study nor the CIS generate diagnostic labels. The latter is concerned only with neurotic symptoms (although psychotic phenomena that become apparent during the interview are noted and recorded in the final section of the Schedule). The diagnosis was made by the interviewer, using clinical judgment and taking into account all the available information.

The decision to use the CIS, rather than an interview such as the Present State Examination (PSE: Wing, Cooper and Sartorius, 1974) which can generate diagnoses, was taken for several reasons:

1. Constraints on time and the large number of subjects made a longer interview impractical.
2. It was predicted that psychotic symptoms would be of low frequency and routine enquiry as to their presence would offend some prisoners and lead to an increased refusal rate.
3. Both interviewers were trained psychiatrists. By using their discretion in conjunction with information already gleaned from the prison record, they could reduce the time spent with prisoners who showed no evidence of mental disorder. When psychotic symptoms were suspected or evident, a clinical psychiatric interview was conducted to determine their nature and content. It was possible to return for a subsequent interview in doubtful or complicated cases.
4. In making a diagnosis, the interview was only one of a number of sources of information. In an attempt to simulate good clinical practice, full weight was given to historical information.
5. For the purposes of the present study, it was less important to assign a precise diagnosis than to identify cases within broad

diagnostic categories, e.g. it was sufficient for our purposes to determine that someone was suffering from a chronic psychotic illness, without being confident about the precise sub-category to which they belonged.

The diagnosis of personality disorder

The factors discussed above apply to all psychiatric diagnoses, and are relevant to all studies which use clinical methods to assign diagnoses. The diagnosis of personality disorder requires further discussion, as it raises three specific problems for a prison survey:

1. The diagnosis of personality disorder is heavily dependent on historical information, particularly from independent sources. In a prison survey, the extent of such available information may be very limited, although it is likely to be available in some cases.
2. The prison environment is abnormal but the structured regime may help to contain some manifestations of an abnormal personality. Therefore, a severe personality disorder, causing enormous problems in the outside world, may not come to notice within prison.
3. Prisoners are, by definition, criminal. Criminality features in many definitions of personality disorder and the diagnosis can be criticized as a relabelling of criminality that has no relevance to psychiatric practice.

In this study, the diagnosis was based on present mental state and behaviour, using historical information when available. Particular note was taken of persistent interpersonal difficulties and social ineptitude, whether this emerged at interview or from other accounts. Behaviour within the penal system is inevitably given added weight, as problems within prison are more likely to be documented than those seen only outside.

While this approach is likely to result in under-diagnosis of personality disorder, it should detect those personalities who are problematic within prison, along with the more severe manifestations of personality disorder outside prison. These biases must be taken into account when comparing the present findings with other studies, and this point will be discussed at greater length when the results are presented.

With regard to the third point, criminal behaviour, self-harm and substance abuse were recorded and coded separately. The presence of these behaviours was not sufficient to justify a diagnosis of personality disorder, unless indicated by additional features of the case. Standardized instruments for diagnosing personality disorder were not used because of the weight which they give to criminality and substance abuse. Many of these instruments are subject to the criticism that they redefine offending in psychiatric terms. A clinical approach, despite problems of reliability, seemed more likely to guarantee relevance to psychiatric practice.

Making a treatment recommendation

In considering the need for psychiatric services, giving a diagnosis to an individual is only the first step. The second and crucial question concerns the treatment that is needed, if any. The resource implications of a patient who can be treated in a general practice setting are quite different from those of a patient requiring a long stay as a hospital inpatient. This difference may not be predictable from diagnosis alone. It depends on additional factors, including the severity of the disorder and the attitude of the individual towards her disorder and towards treatment.

For the purposes of the study, the range of possible treatment recommendations was reduced to the following five:

None
This was automatically applied to subjects with no diagnosis. It was also given to those with a diagnosis but not requiring treatment because of a lack of motivation, unless the nature and degree of mental disorder was such as to require involuntary treatment under the Mental Health Act 1983.

An example would be a woman who was upset by separation from her family. Her symptoms of depression and anxiety were sufficient to warrant a diagnosis but she regarded her feelings as understandable and was coping with them by talking to other women in the prison, who had been very supportive. She had no desire for medical intervention.

Treatment within prison
Treatment which an average GP or psychiatrist could provide on a mainly outpatient basis, e.g. supportive psychotherapy and/or medication. A brief period of special supervision or admission to a prison 'hospital' may be required but with the expectation that most of the sentence could be completed under the normal prison regime without undue risk to the health of the inmate or to staff.

An example would be a woman who had developed anxiety and depression which was preventing her from sleeping, which she was finding intolerable. A doctor, psychologist or probation officer within the prison may provide counselling and the doctor could prescribe medication.

Therapeutic community
The type of contract-based regime found within the prison system at HMP Grendon and outside prison at the Henderson hospital and drug or alcohol rehabilitation centres. Allocation to this category does not require a judgment as to where the treatment is provided but does imply that the inmate is willing to enter into a therapeutic contract.

This category was included as it proved in pilot studies to be a form of treatment with which many drug or alcohol dependent prisoners were familiar. Some inmates were aware of the regime at HMP Grendon but many more had heard of Phoenix House or Clouds House, therapeutic communities which cater for substance abusers. As this form of treatment is rare within the health service, the consequence was that the potential users of the service often knew more about this form of therapy than the medical officers or psychiatrists responsible for their care. For subjects with a diagnosis of drug dependence, this category was used rather loosely to indicate that the subject was motivated to undergo some form of 'psychological' (i.e. not prescribed) treatment aimed at reducing or avoiding future drug use.

Further assessment
This category acknowledges that an accurate diagnosis or treatment decision may be impossible in the prison setting, whether because of lack of information or mixed motives in an environment where 'illness' is one of the few officially sanctioned expressions of distress, anger or discontent. Where there is uncertainty over diagnosis, treatment or motivation, it is assumed that at least the initial stages of assessment would take place in prison, some cases requiring no more than a further interview. The outcome of assessment may range from no treatment to hospital transfer. This category was used only when there was a high degree of suspicion that mental disorder was present and of such a degree as to require intervention.

For example, a woman had been behaving strangely and isolating herself from other people for the last few weeks. At interview, she denied any problems. Her manner was vague but she revealed no other definite symptoms of mental disorder.

Hospital
This term designates inpatient treatment outside the prison system. This category includes all cases requiring compulsory treatment and

inmates willing to accept treatment voluntarily but suffering from psychiatric disorder which cannot be managed adequately or safely in a prison setting. The implication is that the inmate is not correctly placed at present, so an attempt was made to determine how these women reached prison, whether their disorder was recognized within the prison and what arrangements were being made for transfer or management within the prison.

The best example is a woman with schizophrenia, characterized by delusions and hallucinations and attempts at self-harm. Women with severe depression or personality disorders, accompanied by a high risk of self-harm or violence, may also fall into this category.

The need for treatment rests on complex decisions which the researcher was required to make in discussion with the inmate. The aim was to apply current standards of clinical practice in an ideal world, i.e. in the light of present medical knowledge but not constrained by the availability of resources.

There are no specific and generally accepted criteria for assessing treatment needs, so the validity of these assessments is problematic. Although the researchers attempted to apply the standards of current clinical practice, the final decisions were those of particular clinicians and it is likely that other clinicians would arrive at different decisions in many cases. To some extent, this is an unavoidable problem and it can still be claimed that the opinion of one set of clinical researchers is better than none at all. However, an attempt was also made to increase both reliability and validity by replicating the process by which such decisions are reached in the practice of forensic psychiatry, i.e. by discussing each case as a clinical problem, within a multi-disciplinary panel.

The role of the research panel

The approach described above is more efficient than administering a long standardized interview to all subjects. Its disadvantage is that individual symptoms are not routinely recorded and hence cannot be analysed as such. It could be argued that the validity of this method has not been established. It was important to ensure that the decisions taken were not simply the result of biases in the psychiatrists conducting the interviews. One aim of the study is to describe service needs, so it is essential that decisions on diagnosis and management should reflect current clinical opinion within forensic psychiatry.

The research panel was set up to advise on both diagnosis and treatment recommendations. It consisted of the three psychiatrists intimately involved with the study, two senior lecturers in forensic psychiatry (one of whom was also a special hospital consultant), two psychologists working in the forensic field, a nurse from a regional secure unit and a probation officer. At monthly meetings, this group considered all cases suspected of suffering from severe psychiatric disorder in addition to any other cases felt by the interviewers to present diagnostic or management problems. Information considered included interview data and all available reports, e.g. past psychiatric and social inquiry reports, the disciplinary record in prison and detailed statements regarding the index offence.

All cases selected as above were debated by the group and a consensus was usually reached. In the rare cases of unresolved disagreement, the interviewer made the final ratings in the light of the points raised during discussion.

Reliability

Inter-rater reliability for other items on the questionnaire was checked in a pilot study when twenty volunteer prisoners were interviewed by both interviewers on separate occasions. The information obtained was used to compile the coding manual. A few items were dropped from the questionnaire when it proved impossible to establish adequate reliability.

During the course of the study, a number of measures were taken to ensure that reliability was maintained. The interviewers discussed many of the cases informally or at larger meetings of the research steering committee (see below). Periodically, a small number of prisoners (approximately 5 per cent of female subjects and 2 per cent of males) was interviewed by both interviewers and results compared.

Statistical techniques

The samples of women and men represent 25 per cent and 5 per cent of their respective prison populations. SPSS–X is used for most analyses. In addition, confidence intervals for proportions, differences and odds ratios are given where appropriate.

Confidence intervals

Confidence intervals provide an estimate of population values derived from measurements on the sample. The 95 per cent confidence interval is used throughout this project. It indicates that the chances of the true value of the population statistic concerned falling outside the range of values given is not greater than 5 per cent. Confidence intervals can be calculated for many statistics, including proportions, the difference between proportions and odds ratios.

Example: It will be shown below that psychosis is diagnosed in 1.6 per cent of the women in the study sample. It cannot be assumed that psychosis will be found in exactly 1.6 per cent of all women serving a prison sentence, i.e. in the prison population. By calculating a confidence interval, we can be 95 per cent certain that the prevalence of psychosis in the female prison population lies in the range 0.3 per cent to 2.9 per cent.

The finite population correction
The formula for calculating the confidence interval for a proportion or percentage takes account of sample size but makes no reference to the population size; the assumption is that the population is infinite or of unknown size. When a population is of known and finite size it is possible to use a finite population correction. In practice, this is only worthwhile when the sample is large relative to the total population – a condition which applies to the sample of the female prison population in the present study. The population estimate of 1048 (see above) has been used to calculate the finite population correction for UK resident women serving a prison sentence; in fact, the effect of substituting the original population figure of 1229 is negligible.

Odds ratios

Odds ratios are used as a test for the difference between proportions in a 2×2 table, as an alternative to the Chi^2 test. The odds ratio is an estimate of relative risk. A result of 1.0 indicates that the proportion of each of two groups with a particular condition is identical. The greater the deviation from 1.0, the greater the difference in proportions. Unlike the Chi^2 test, this statistic describes the magnitude of the difference between proportions; with large samples, a Chi^2 test may show statistical significance when the magnitude of the difference between proportions is quite small. Chi^2 tests are used occasionally, when the magnitude of a difference is of less importance.

Confidence intervals are calculated for odds ratios. If the 95 per cent confidence interval for the population odds ratio does not include the value 1.0, the null hypothesis (that the population proportions are in fact equal) can be rejected and it can be assumed that a Chi^2 test would have been significant at the $p < 0.05$ level or higher. In the tables which follow, odds ratios for which the 95 per cent confidence interval for the population odds ratio does not include the value 1.0 are given.

Chapter summary

A sample of women, comprising 25 per cent of the sentenced prison population, was selected from four prisons. It was established that the sample was representative of the female prison population, in terms of offence type, sentence length and ethnic origin. The ethnic composition of the female prison population differs from that of male prisoners, due to the presence of a high proportion of drug smugglers from other countries. This will be taken into account in the analysis of results, and women from overseas will be described separately.

Using data from prison and hospital records, and a semi-structured psychiatric interview, each woman was assigned a psychiatric diagnosis when appropriate, and a decision was made as to the most appropriate treatment in each case. A 5 per cent sample of the male sentenced population formed a comparison group, in whom identical diagnostic methods were used. In the following chapters, the results of this survey are presented.

4

Results I: Violence by women in prison

Introduction

The ultimate goal of the prison survey was to estimate the need for medical services, so its first concern was with the prevalence of psychiatric disorders. However, important as these figures are, they are a tiny part of a full description of female prisoners. Many gender differences are better described in criminological or behavioural terms, than by reference to psychiatric diagnosis. The first results to be presented are part of this broader picture.

The findings in this chapter are all concerned with violence, in its broadest sense. They include violence towards others, towards the self, and violence directed at authority. The chapter begins with a description of the violent offending of a minority of women, including those who were imprisoned because they had killed. The main concern is with past behaviour, the offending which led to imprisonment. Offences against prison discipline are then used to illustrate behaviour within the institution, demonstrating that such behaviour is often shaped and constrained by the prison environment. A description of self-harming behaviour reveals continuities between violence to others and to the self. The turning inward of violent impulses is thought to be characteristic of women, but the gender comparison shows that, in prison, this difference is not clear cut.

The emphasis is on women as perpetrators of violence. Of course, in any overall view of violence, it is true that women are more often victims than perpetrators. This is no reason to avoid describing those women who do commit violent acts, whether before they come to prison or while they are in custody. There is a particular need for data in this area. Violence by women has been neglected by researchers, although there has been much speculation about its nature and causes – including speculation that women's violence is always evidence of psychiatric problems. The findings of the present study undermine some of this speculation.

Violent offending

Violent offending is much less common in women than in men. About one in five of all offenders convicted or cautioned in England and Wales is a woman but, for violence against the person, this ratio falls to 1:8. Even this figure understates the gender difference, as the offences committed by men are more often of a serious nature. As a result, the gender ratio in sentenced prisoners is greatly skewed. In 1988, 23 per cent of male sentenced prisoners were serving a sentence for violence, compared to 21 per cent of female prisoners. The absolute numbers were 247 and 8586 respectively, a ratio of one woman to thirty-five men.

The samples used in the present study reflected the prison statistics almost exactly. Of the 258 women interviewed, fifty-three (21 per cent) were serving a sentence for violent offences, compared to 22 per cent of the male sample.

These figures are misleading, as they take no account of previous violent offences, or of the seriousness of the offence. For example, one of the women and several men had been sentenced to imprisonment for affray, having become involved in a large crowd confronting police after a rowdy party. The offence is violent, but does the conviction make the offender a violent person? Conversely, a woman serving her present sentence for burglary (motivated by the desire to buy alcohol) had a record of four previous convictions for violence; it seemed unreasonable to exclude her from any group of violent women.

For this reason, a sub-group of 'violent offenders' was defined, as prisoners with a record of three or more convictions for minor violence or a single conviction for severe, life-threatening violence. This definition is similar to that used in the Criminal Profile (Gunn and Robertson, 1976), so the information was already coded. Using this definition, forty-three women (17 per cent) and 439 men (25 per cent) were labelled as violent offenders, emphasizing again the greater prevalence of serious violent offending among male inmates.

Further gender differences emerge when these violent offenders are described in more detail. Table 4.1 shows the current or index offences (which may not be violent, if there is a record of previous violent offending).

The pattern of current offences is quite different, with murder or attempted murder accounting for almost two-thirds of the violent women but only one-third of the violent men. Of the twenty-six women sentenced for murder or attempted murder, twenty-two were serving life for murder and none of these women had ever been in prison before. By contrast, the 120 men serving life for murder

Table 4.1 Violent offenders by current offence and by gender

	Women		Men	
	n	%	n	%
Murder or attempt	26	60	154	35
Assault/wounding	6	14	95	22
Theft/robbery	6	14	141	32
Sex offences	–	–	22	5
Arson	3	7	6	1
Other	2	5	21	5
Total	43	100	439	100

included thirty-nine (33 per cent) who had served at least two previous sentences. None of the women sentenced for murder had a significant criminal record before the index offence, whereas an extensive criminal history was common in equivalent men.

The twenty-two women sentenced for murder were compared to twenty-two similar men, matched for age at time of sentence. The relationship of offender and victim is summarized in Table 4.2.

Further information was obtained from trial reports. Although the prisoners described in this table had been sentenced for the same offence, there were enormous differences in the behaviour and social context leading to the conviction. The circumstances of the killings in the case of the women were a triangular sexual relationship (ten cases), domestic violence (eight), financial gain (three) and terrorism (one). The circumstances of the killings by men were disputes with acquaintances (ten cases), financial gain (five), sexual (four), domestic violence (two) and a triangular sexual relationship (one).

It would be wrong to make too many generalizations from these figures, as the number of cases is small (although it should be remembered that the twenty-two women represent one-quarter of all females serving a sentence for murder in England and Wales at

Table 4.2 The relationship of victim to offender in women and men serving a life sentence for murder

Women N = 22		Men N = 22	
Husband/cohabitee	16	Stranger	11
Other family member	2	Male acquaintance	9
Dependent adult	2	Wife/cohabitee	2
Stranger	2		

the time of the study). Further statistical analysis is not possible. However, despite these limitations, a number of points can be made.

First, gender differences are obvious but not absolute. It was unusual for women to kill a stranger, but two had done so. One man had killed a sexual partner in the context of a relationship triangle, although this was a much commoner scenario in women. Disputes and arguments (excluding those with a spouse) were common in the male homicides but had not been important in any of the female cases. Financial gain appeared to be an important motive in similar numbers of men and women.

Second, it is not easy to explain the gender differences in simple psychological or moral terms. The commonest situation in the male homicides was an argument that got out of hand. For example, one man in his twenties was walking home from the pub on Friday night and became involved in an argument with a man of similar age, over some past insult. Similar scenes are played out in many towns on most weekends. On this occasion, one man had a knife and the other was killed. The circumstances will be familiar to any psychiatrist with experience of preparing 'routine' reports in murder cases, many of which cases are surprisingly banal. One feels there should be more significance to events in which a person's life was lost but it is difficult to find much.

By contrast, none of the murders by women had this element of chance violence and almost random selection of victim. In many cases, there were clear and understandable reasons for wanting the victim dead. In most, there had been an intimate relationship between the woman and the victim, although this was often over by the time of the offence. The following case vignettes illustrate some of these points.

Women sentenced for murder: case vignettes

Case no. 437
A 33-year-old woman, with no previous convictions, was married to the landlord of a country pub, with two children below the age of ten years. She had become bored with her husband and began an affair with a younger man who came to work as a barman at the pub. Together, they plotted to kill her husband. Her lover carried out the fatal attack, attempting to make it look like a break-in. At interview, she was three years into her life sentence. There was no evidence of previous psychiatric disorder but she was now suffering from a reactive depression, brought on by news of the length of time she would have to spend in prison before becoming

due for consideration of parole. Treatment, in the form of supportive psychotherapy, was being given by a visiting psychiatrist at the prison. At interview, she accepted that she had plotted the death of her husband but believed her tariff to be unreasonable, as her lover had been the prime mover in the plot, in addition to having carried out the act.

Case no. 438

A 40-year-old woman with no previous convictions had plotted the killing of her husband, along with two other men, both of whom were her lovers. The two men carried out the act, after she lured the victim to a lonely place. At interview, ten years into her life sentence, she presented as a woman with a very odd personality, whose account was dominated by hypochondriacal complaints and a list of injustices perpetrated against her by the criminal justice system. She had given many accounts of the offence, earning her the label of 'pathological liar', and now claimed that it was a bizarre suicide pact. She was given diagnoses of personality disorder and hypochondriasis and the researchers judged her to be in need of further psychiatric assessment, because of her bizarre presentation.

Case no. 462

A 29-year-old woman of Asian origin, with no previous convictions, plotted with her lover to kill her husband, a much older man. She believed he would oppose a divorce, as would her extended family. The lover carried out the act. At interview, six years into her sentence, she had symptoms of depression for which she was receiving treatment from the visiting psychiatrist at the prison.

Case no. 387

A 41-year-old woman had an unhappy second marriage. Her new husband resented her past and refused to speak to her two daughters from the first marriage. As the situation deteriorated, she began an affair and persuaded the man to kill her husband, partly so that they could be together but also for a payment of several thousand pounds. At interview, one year into her life sentence, she admitted to heavy drinking throughout the period leading up to the offence but showed no other evidence of psychiatric disorder.

Case no. 2266

A 47-year-old woman had been the part-owner of a business with her husband, both having connections with organized crime. They divorced, with much acrimony over money, particularly the division of business interests. She paid a man to enter her ex-husband's house as an intruder and kill him. In circumstances which are not clear, she accompanied the intruder and struck the fatal blow. At interview, three years into her life sentence, there was no evidence of psychiatric disorder.

Case no. 383

A 16 year old who, together with her boyfriend, attacked her ex-boyfriend and killed him. The reasons are not entirely clear, although they included jealousy on the part of her new boyfriend and complaints from her about the way her ex-boyfriend had treated her and 'dumped' her. At interview, six years into her life sentence, there was no evidence of past or present psychiatric disorder.

Case no. 2265

A 17-year-old woman was having an affair with an older, married man. Together, they plotted the killing of his wife, an action which the man carried out. The extent of her involvement in events was disputed, although she had been found guilty of murder. At interview, eighteen months into her life sentence, she was suffering from an eating disorder that had been present since the age of 15 years.

Case no. 2259

A 42-year-old woman had been looking after her disabled, wheelchair-bound husband for several years and grew to resent his demands for attention, and his inability to give much in return. Together with three male acquaintances, she planned and carried out his murder. At interview, eighteen months into her life sentence, she showed no evidence of psychiatric disorder.

Case no. 2252

A 22-year-old woman, whose common law husband was serving a prison sentence, began an affair with another man. They planned and carried out the murder of her husband when he was released. The victim was a violent man who had abused his wife on many occasions, and both she and her lover feared he would kill them if he were allowed to discover their affair, or if she tried to leave him. At interview, one year into her life sentence, there was no evidence of psychiatric disorder.

Comments

Each case has highly individual features and it is dangerous to draw too many general conclusions. Nevertheless, several points stand out. The first is the involvement of men in most of the killings, even in cases where there is no doubt that the woman was the instigator of the offence. The victim often stood in the way of the woman's new relationship, so a motive was truly shared, but the man was more physically capable of carrying out the killing, which the couple planned together.

The men involved in these murders also received life sentences. In other words, the motives and circumstances of the women who had murdered were not unique to women. At least as many men, if not more, are likely to be serving sentences for offences committed in similar circumstances. The difference is that these men are vastly outnumbered by those who have killed for other reasons, in the course of petty quarrels or robberies. For all practical purposes, this latter group does not exist within the female prison population. The commonest type of violent offending among men is one in which women rarely become involved.

Theoretical frameworks

There have been few systematic attempts to explain homicides by women. Hartmann (1977) described thirteen middle-class Victorian women who were found guilty of murder and concluded that their acts of violence represented extreme solutions to problems faced by many of their contemporaries. The cases described above fit this framework reasonably well. In most, the act of killing was planned to be instrumental in achieving a desired result, whether elimination of an inconvenient spouse or financial gain. Unfortunately, the explanation is incomplete and begs the question: what distinguishes those women who kill from those with similar problems who attempt other solutions, or continue to tolerate the problem? The triangular relationship, for example, is extremely common, yet rarely results in homicide.

Jones (1991), in her book *Women Who Kill*, provides a historical survey of homicides by women. She attempts to link the crimes to the history of women in America, showing how women's motives for killing alter, as their roles within society are also changed. She argues convincingly that gender differences in homicide can never be fully understood in isolation from the wider context of gender roles within society. Jones also makes the point that gender differences in violence are not absolute and that many men murder those they love, just as some women kill for financial reasons, in the heat of an argument or, occasionally, for pleasure. Like Hartmann, Jones's emphasis is on the continuity between women who kill and the majority who do not. It leaves the reader still curious about those factors which may serve to distinguish the tiny minority who behave violently.

Prolonged victimization, by a violent husband, has received much publicity as a factor that may contribute to homicidal behaviour by women. There is concern that the law, administered mainly by men, does not recognize the plight of women subjected to repeated violence within the home, a tiny minority of whom respond by killing their partner/persecutor. The most celebrated individual case is that of Sara Thornton, whose defence counsel argued without success that a long history of violence by her husband amounted to provocation, even when the defendant delayed her attack until her victim was asleep, for reasons of self-preservation (*R. v. Thornton, 1991*). The Thornton case is described in detail in Nadel (1993). At the time of writing, Thornton remains in prison but a miscarriage of justice was confirmed in the case of *R. v. Alawalia*, who had killed her violent husband and later had her murder conviction overturned on appeal.

The publicity generated by these cases led the Home Office to release figures which suggest that they do not represent a general trend for the law to discriminate against women (Table 4.3). Between 1982 and 1989, 177 women and 785 men in England and Wales were charged with the killing of their partner or former partner (a total of 962 offences, with women accounting for 18 per cent). The charge was reduced to manslaughter in 7 per cent of the women and 4 per cent of the men. At trial, 22 per cent of the women were acquitted (or found unfit to plead), compared to 5 per cent of men.

The murder convictions result in mandatory life sentences. The punishments imposed on those convicted of other offences are shown in Table 4.4.

The actual sentencing differences are even greater, as the average sentence length does not take account of the life sentences given to some men. None of the women convicted of manslaughter were given a life sentence.

Table 4.3 The convictions recorded against persons found guilty of killing their spouse in England and Wales, 1982–1989, by gender

	Women N = 137	Men N = 748
Convicted of murder	20%	37%
Manslaughter on grounds of diminished responsibility	26%	32%
Manslaughter on other grounds (usually provocation)	51%	30%
Others	3%	1%

Table 4.4 Punishment imposed on persons convicted of manslaughter of a spouse in England and Wales, 1986–1989, by gender

	Women	Men
Imprisonment	45%	70%
Average sentence length	33 m.	59 m.
Probation or suspended sentence	27%	8%

It must be emphasized that these figures are of no relevance to an individual case, where there may or may not have been a miscarriage of justice. However, these figures do suggest that discrimination by the courts against women who have killed violent husbands is not a factor in the cases of most women imprisoned for murder.

In the present study, domestic violence was the second most common setting in women convicted of murder. It would be presumptuous and quite wrong for the researchers to make any judgment as to whether the verdict of the court was correct in these cases. Nevertheless, the figures do show enormous qualitative differences between male and female violence, and it is important for the courts (and doctors writing reports) to be aware of these differences, if miscarriages of justice are to be avoided.

Disciplinary offences within prison

History and meaning

Women's behaviour inside prison has long fascinated (predominantly male) authors. Dobash, Dobash and Gutteridge (1986) review the historical development of prison regimes for women, drawing attention to the conventional wisdom that women, while less violent outside prison, behave more violently than men once locked up. Mayhew and Binny (1862) suggested that women in London prisons were notorious for a high level of disturbed behaviour in the mid-nineteenth century. As the chaplain of Brixton prison put it: 'Violence of temper is one great evil with female prisoners: they are so easily excited, and so subject to sudden impulses, that it is very painful to consider what misery they bring upon themselves, owing to the influence of bad temper.' The 'misery they bring upon themselves' was a euphemism for punishment by the prison authorities. Figures for recorded punishments showed that the average per prisoner per year was less than two for men, but three and a half for women. Quinton (1910) suggested that little had changed and commented on the need for higher staffing levels to manage the women's section of Millbank prison.

The review by Dobash et al. makes the point that disturbed behaviour by women can be seen in two ways. It may be regarded as evidence of women's inherent emotionality and irrationality, once released from the constraints of society. This explanation serves a clear moral purpose, reinforcing the need for tight social controls on the behaviour of women. On the other hand, the behaviour can be seen as self-expression and a rebellion against the prison regime, which was undoubtedly oppressive. This explanation has appeal for feminists, portraying women in prison as rebels, fighting against an unjust and patriarchal system.

A third perspective should also be considered: that some rule-breaking by women in prison arises from mental disorder. For our purposes, it does not matter if mental disorder accounts for only a small proportion of disciplinary problems. The point is that it requires identification, and a different form of management. While some critics are content to lump psychiatry together with other institutional responses, as another form of social control, its significance for an individual woman may be enormous. When a prisoner is charged with an offence against prison discipline, a doctor within the prison has the power to declare that she is 'unfit for adjudication'. Women (or men) who are fit for adjudication appear before the Governor or Board of Visitors to have their case heard and, if found guilty, punishment imposed. Those who are unfit are shifted onto a different track and assessed for medical treatment, a process which may lead to transfer from prison and into hospital.

This process is important because of the disastrous consequences if it does not operate effectively. One of the criticisms of Holloway in the mid-1970s was the high rate of disciplinary offences, and the impression that doctors had abandoned their role in deciding who was fit for adjudication and who was not. Women with serious mental illness were being punished for behaviour over which they had little control. In cases of severe mental disorder, it is possible to progress rapidly through a prison's disciplinary system, to the point where an inmate is held in inhumane conditions and deprived of all privileges. The acceptable face of 'medicalization' is that it allows intervention in this process, to provide treatment rather than punishment.

Recent statistics and the survey results

Before considering further explanations for women's offences against prison discipline, it is necessary to look at recent statistics. It remains the case that women within prison are punished for disciplinary offences more often than men. The rate of offences per 100 prison population for all female establishments in 1985 was 335, whereas for men it was 160. These figures are not so different from those of a hundred years earlier. As in the nineteenth century, the majority of disciplinary offences today concern acts of disobedience or disrespect.

Table 4.5 shows data on disciplinary offences which were collected in the course of the prison survey.

About a third of all women had a disciplinary offence recorded against them during the three months prior to interview, compared to only one-fifth of men. As the statistical tests show, this difference is unlikely to be due to chance.

When disciplinary records are examined in more detail, eight women (3 per cent of the sample) had a record of five or more disciplinary offences during the same period, compared to nineteen men (1 per cent). A greater proportion of women had offended against prison discipline, and a greater proportion were frequent offenders. So far, the conventional wisdom is borne out by the findings.

Most rule violations by prisoners are dealt with by means of small fines or loss of privileges. However, Table 4.5 also shows gender differences for two disciplinary measures used against prisoners who have committed more serious offences. Transfer to another prison is most often used as a sanction against inmates who

Table 4.5 Disciplinary record during current imprisonment, by gender

	Women N = 262			Men N = 1751			Odds ratio	
	n	%	N/A	n	%	N/A	OR	95% c.i.
Disciplinary offence in last 3 months	83	32	1	361	21	10	1.8	1.4–2.4
Disciplinary transfer during present sentence	36	14	5	365	22	89	0.5	0.2–0.9
Any loss of remission during present sentence	19	7	1	651	37	7	0.1	0.08–0.2

Notes: N/A = information not available. Percentages and odds ratios in this table are calculated after subtracting N/A from N. c.i = confidence interval.

abscond, or repeatedly breach prison rules. It usually involves a move to a higher level of security, e.g. from an open to a closed institution. While 14 per cent of women had been transferred for disciplinary reasons during their current imprisonment, the figure for men was significantly higher at 22 per cent.

The most serious punishment used by the prison, without resorting to the courts, is loss of remission. This is a severe and controversial punishment, as it prolongs the time spent in custody, thereby effectively increasing the sentence imposed by the court. Only 7 per cent of the women in our sample had ever lost any remission during their present sentence, compared to 37 per cent of men. This is a highly significant difference, even allowing for the fact that men may be serving longer sentences.

Discussion

A clear picture emerges from these figures. They do not suggest that women in prison are generally more disturbed or violent than men. Rather, the management of men within prison presents more serious disciplinary problems, while women are more likely to be disciplined for trivial offences. This point has been made by Carlen (1985), who cites it as evidence that prison regimes for women are more oppressive than those for men, and require a higher standard of behaviour. Mandaraka-Sheppard (1986) argues that aggressive offences by female prisoners can best be explained as 'a function of their response to the particular negative aspects of institutions'. Her study of three open and three closed female prisons found institutional rather than individual characteristics to be more important in accounting for violent incidents. She criticized the trivial nature of many recorded offences, the vague nature of some rules and the consequent inconsistency in their application. It has also been suggested that the nature and attitudes of female prison officers are partly responsible for the high numbers of trivial offences recorded in women's prisons (Kozuba-Kozubska and Turrel, 1978).

All these factors may be important. The comparison with male prisoners is complicated by the difference in regimes. In the prisons visited for the present study, women had more physical freedom within prison than men with a similar security classification. The long lock-up periods that characterize male prisons, even those for long-term inmates (King and McDermott, 1989), are much less common in female prisons.

Holloway, at the time of the study, operated a 'free flow' system whereby, for certain specified periods, all doors on the prison's main corridor were unlocked. Women moved freely between the various activities and locations. This contrasted with the usual method of moving prisoners, in groups with escorting officers. 'Free flow' was generally preferred by the women and gave them greater trust. At the same time, it created a new possibility for breaching regulations, by failing to get to the right place at the right time – an option that was not available to the prisoner with a traditional escort.

In the same way, Styal provided considerable freedom of movement within a secure perimeter for the majority of its inmates. This allowed more interaction with prison officers and increased the potential for transgressing minor prison rules. For similar reasons, many inmates stated that they found 'open' prisons more oppressive. The increased freedom meant increased regulation of behaviour and some chose to return to closed prisons. Mandaraka-Sheppard's description of an open prison regime brings out the paradox whereby (conditional) freedom of movement increases the opportunities to breach regulations.

These points have important consequences for mentally disordered prisoners. Most women (and men) who break prison rules are not mentally disordered but a minority of inmates will repeatedly breach rules because of their serious mental illness. The more open the prison regime, the more difficult it may be for mentally disordered inmates to conform. For this reason, a history of mental illness may preclude transfer to open prison conditions. Within a generally lax prison regime, the mentally ill prisoner may find her freedom greatly restricted, as was the case at Styal. This problem will be discussed in more detail in the next chapter. For the present, it should be noted that improvements in the regime which benefit most prisoners may not help those with serious mental disorders, further reinforcing the principle that the mentally ill should be looked after in the health service, not prison.

Violence to the self: deliberate self-harm

Criticism of prison conditions often highlights deliberate self-harm (DSH) by prisoners. Most people are frightened and repelled by the prospect of inflicting injury upon themselves, so assume that a regime must be truly inhuman if inmates are driven to mutilate or even destroy themselves. A high rate of self-injury was one of the factors that precipitated action to reform the regime at Holloway in the mid-1970s (Moorehead, 1985).

It is conventional to separate self-injury into suicidal behaviour (aimed at ending life) and other forms of deliberate self-harm (also termed parasuicide) which are assumed to be manifestations of psychological distress, rather than attempts at suicide. The distinction is valid in so far as there are differences between the two populations, with parasuicides tending to be younger and female. However, there is much overlap between the two groups. In the present study, the nature of self-harming behaviour was recorded independently of an assessment of suicidal intent.

The survey results

At interview, women were asked whether they had ever tried to kill themselves, had ever deliberately harmed themselves or taken an overdose. In all, eighty-two women (32 per cent) reported deliberate self-harm on at least one occasion in the past and thirty-one (12 per cent) said they had harmed themselves more than once. These figures, suggesting that a third of all female prisoners have a history of self-harm, are higher than those for male prisoners, of whom 297 (17 per cent) reported self-injury, including 131 (8 per cent) who had self-harmed on more than one occasion. Table 4.6 provides a more detailed comparison of deliberate self-harm in male and female prisoners.

The table shows that the comparison with men is more complex than it first appears. Women do have a higher prevalence of deliberate self-harm but only when self-harm *outside* prison is taken into account. When DSH in custody is considered alone, 5 per cent of female and male prisoners report this behaviour at least once during the present or previous times in custody. These findings are reinforced when method of DSH is considered. Table 4.7 shows that the only significant gender difference is the higher prevalence of overdoses in women.

The picture emerging from these two tables is that women in prison have a higher lifetime prevalence of deliberate self-harm than men, but clear-cut differences are limited to overdosing outside

Table 4.6 The lifetime prevalence of deliberate self-harm (DSH) in sentenced prisoners, by location and gender

| | *Women* N = 258 | | *Men* N = 1748 | | *Odds ratio* | *95% c.i.* |
	n	%	n	%		
Any past DSH	82	32	297	17	2.3	1.7–3.0
DSH in custody	13	5	81	5	1.0	–
DSH outside	78	30	236	14	2.8	2.1–3.7

Table 4.7 Lifetime prevalence of deliberate self-harm in sentenced prisoners, by method of self-harm and by gender

	Women N = 258		Men N = 1748		Odds ratio	95% c.i.
	n	%	n	%		
Overdose	64	25	156	97	3.4	2.4–4.7
Cutting	25	10	151	9	1.1	0.7–1.8
Hanging	7	3	22	1	2.2	0.9–5.2
Other	6	2	48	3	0.8	0.4–2.0

Notes: An individual may report more than one method of deliberate self-harm. 'Other methods' consisted of jumping from high places, fire (reported by seven men and two women) and gassing.

prison. Prevalence rates for DSH during the current period in custody are the same for both sexes. Also, the lifetime prevalence of DSH by cutting in women is 10 per cent (the 95 per cent confidence interval is 7–13 per cent), not significantly different to the 9 per cent found in male prisoners.

Previous studies of self-harm in prisoners

Both the male and female rates for self-cutting are similar to the 8 per cent prevalence rate found in Wilkins and Coid's (1991) survey of remanded women. Taken together, the most surprising aspect of these figures is the high prevalence of self-cutting in male prisoners, which has received little attention in the literature.

A more detailed analysis of the twenty-five women who had cut themselves found that thirteen had done so on more than one occasion. Seventeen of these women reported DSH only outside prison. A history of self-cutting showed a positive association with several psychological or social variables. Women who have been in care in childhood have a lifetime prevalence of DSH of 19 per cent, compared to 8 per cent in those who have not (the odds ratio is 2.9, with a 95 per cent confidence interval of 1.2–6.9). The prevalence in the thirty-six women with a drink problem is 22 per cent, compared to 8 per cent in other inmates (o.r. = 3.5, 95 per cent c.i. = 1.4–8.9), and the prevalence in women with a history of violent offending (defined as at least one conviction for violence against the person) is 15 per cent, compared to 6 per cent in those with no history of violence (o.r. = 2.6, 95 per cent c.i. = 1.1–6.0). Drug-dependent women have a lifetime prevalence of self-cutting of 13 per cent, compared to 9 per cent in non-dependent women, a difference which is not statistically significant. (See Maden, 1992 for further details.)

Wilkins and Coid found self-mutilation in female prisoners to be associated with violence and arson, and regarded it as an indicator of severe psychopathology. The Holloway study by Cookson found an association of self-injury with violent offending. The present study confirms that women with a history of self-cutting are more likely to have been in care, to have a history of violent offending and a drink problem but finds no significant association with drug dependence. The same pattern of associations was found in the male sample, suggesting that similar psychological mechanisms may be operating in both men and women who self-injure.

Liebling's (1992) survey of young offenders found DSH in women to be associated with (among other factors) a history of sexual abuse and violence within the home. Compared to other women in prison, those who self-harmed were more likely to have a history of violent offending or arson. Alcohol problems also showed a strong association with self-harm.

In considering responses to self-harm, Liebling also found that women were more likely than similar men to have been referred for psychiatric treatment, and were more likely to turn to staff for help with personal problems, whereas male prisoners would keep problems to themselves. She comments that this may represent differences in the willingness of staff to respond, as much as it represents any gender differences in individual psychology.

In summing up the differences between male and female young offenders who self-harm, Liebling characterizes the women as showing more characteristics of the 'pathology model', particularly in their histories of violence. This is consistent with the findings of the present study.

Chapter summary

There are qualitative differences between women and men in prison, in respect of their violent offending, disciplinary offences within prison, and self-harming behaviour. These differences are not easily reduced to a single dimension of psychopathology; they cannot be explained away by suggesting that women are simply more 'disturbed' than men. In the case of deliberate self-harm, the extent of pathology in male prisoners has not been given sufficient recognition, and their rates of self-cutting are comparable to those in female prisoners. There is some evidence to suggest that professionals are more willing to acknowledge and respond to evidence of psychological disturbance in women, whereas it is ignored or treated as a disciplinary matter when it occurs in male prisoners.

5

Results II: The prevalence of psychiatric disorders

Introduction

The results described in this chapter form the crux of the prison survey. The prevalence of psychiatric disorders is described separately for women who are normally resident in the United Kingdom, and for women who are resident overseas. Rates of psychiatric disorder in female and male inmates are compared, using only prisoners who are resident in the UK, for the reasons given in Chapter 3. An attempt is then made to relate gender differences in prisoners to data on psychiatric disorders in the wider community.

The final section considers utilization of services. Women in prison are shown to receive more psychiatric treatment than men. This finding is only partly explained by the greater prevalence of psychiatric disorder in women, and two other factors are explored: the readiness of the prison to provide treatment to prisoners, and the attitudes of male and female prisoners towards treatment. The chapter ends with a discussion of links between psychiatric diagnosis and ethnic origin.

A note on diagnosis and labelling

Diagnostic labels are central to the following discussion, so it is necessary to be aware of the uses and limitations of diagnosis in psychiatry. The title of this chapter refers to disorders in the plural. This is a deliberate attempt to get away from the practice of lumping together all persons given a diagnosis under the heading of psychiatric 'cases'. There is little common ground between a woman with mild depression following unpleasant life events, and another with a twenty-year history of recurrent hospitalization for schizophrenia. Statements such as 'one-third of prisoners are psychiatric cases' have little meaning unless the diagnosis (the type of mental disorder) is specified. Even when a diagnosis is given, the need for treatment is not always clear, and this problem is dealt with in the next chapter.

A further problem with diagnostic labels is that the reader may make various unjustified inferences. For example, it may be assumed that a particular label implies that a person is incompetent, not responsible for her actions, or in need of a particular treatment. All observers have their own prejudices and expectations, and the result is confusion. For the purposes of the present study, diagnostic labelling followed the standard practice of psychiatric epidemiology. An accepted diagnostic system was used, and a diagnostic label meant only that the behaviours and experiences of an individual fitted the description found in this scheme. Decisions about treatment were then taken as a separate, explicit step.

These points can be illustrated by the example of drug dependence. There is plenty of room for futile argument about whether or not this is a disease, a behaviour disorder or a habit. The debate can be side-stepped by the following definition:

> A state, psychic and sometimes also physical, resulting from a drug, characterized by behavioural and other responses that always include a compulsion to take a drug on a continuous or periodic basis in order to experience its psychic effects, and sometimes to avoid the discomfort of its absence.
>
> (World Health Organization, 1978: 42)

Of course, there is still room for discretion, and argument about the severity of symptoms, but this definition makes no assumptions about causation, and does not imply a particular treatment (or any treatment, for that matter). Throughout this study, diagnostic labels will be used in a similar way, as a convenient shorthand for patterns of behaviour and experience.

The system of classification used in the study was the World Health Organization's (1978) International Classification of Diseases, 9th revision (ICD9). In order to simplify analysis, some of the categories in ICD9 were modified and combined as follows:

Psychosis includes categories 295 (schizophrenia), 296 (affective psychoses) and 297 (paranoid states).
Neurotic disorder includes categories 300 (here referred to as neurosis) and 309 (adjustment reaction).
Personality disorder is category 301.
Substance abuse includes categories 303 (alcohol dependence), 304 (drug dependence) and 305 (non-dependent abuse of drugs).
Organic disorder includes categories 317 (here referred to as mental handicap) and 345 (epilepsy). The latter is included as a neurological disorder that may have psychiatric complications.
Sexual deviation is category 302.

Any other uses of diagnostic terms will be explained at the appropriate point in the text.

Psychiatric disorder in women serving a prison sentence

After excluding all overseas residents from the analysis, 147 women (57 per cent) were given at least one diagnosis. Thirty-nine of these women also received a second diagnosis and four were given a third. The prevalence of all these diagnoses (i.e. including second and third diagnoses) is shown in Table 5.1.

The decision as to which diagnosis is primary is somewhat arbitrary. The commonest second diagnosis was personality disorder and most of the cases with multiple diagnoses consist of permutations of substance abuse, personality disorder and neurosis. Five women with mild mental handicap and one woman with epilepsy also received a second diagnosis of personality disorder. In the following account, all references will be to the total prevalence rates, i.e. including second and third diagnoses. The total, including subjects with no diagnosis, will therefore exceed 100 per cent, as it does in Table 5.1.

At first glance, the results are striking. Almost two-thirds of all women serving a prison sentence can be given a psychiatric diagnosis. However, most of these diagnoses are for substance abuse, personality disorder and neuroses. Psychosis was found in less than 2 per cent of the women interviewed.

Particular diagnoses and their implications for treatment will be discussed later. For now, it should be noted that the treatment implications of the common diagnoses are unclear. Most people would agree that schizophrenia warrants a detailed psychiatric

Table 5.1 Female sentenced prisoners: frequency of psychiatric diagnoses, with confidence intervals

Diagnosis	n	%	95% c.i.
Psychosis	4	1.6	0.3–2.9
Neurotic disorder	40	16	12.2–19.8
Personality disorder	46	18	13.9–22.1
Alcohol abuse/dependence	24	9	5.9–12.1
Drug abuse/dependence	67	26	21.3–30.7
Mental handicap	6	2.3	0.7–3.9
Other disorders	2	0.8	–
Epilepsy	1	0.4	–
No diagnosis	111	43	–

assessment with a view to treatment with medication, possibly in hospital, but this is not necessarily so with drug dependence or neurotic disorders. The role of treatment in these cases depends on many factors, not least the views and wishes of the individual concerned. The statement that two-thirds of prisoners are psychiatric cases, in a technical sense, does not mean that these women would regard themselves as patients, nor does it imply that they would be regarded as such by other people or the courts.

The broad pattern of diagnoses is consistent with the results of most surveys of sentenced prisoners (female and male), with low rates of psychosis and high rates of personality disorder, substance abuse and neurosis. A similar pattern was found in the men interviewed for the present study, whose diagnoses are reported below and described in more detail elsewhere (Gunn, Maden and Swinton, 1991).

The concluding parts of this book are concerned with the study's implications for psychiatry and for psychiatric services. It would be premature to discuss these matters at length here but one important implication for the psychiatrist in prison is obvious: a doctor whose interests are restricted to the psychoses is not suitable for the job. It is crucial that a psychiatric service within prison deals efficiently with the small proportion of psychotic prisoners, but a much broader range of services is also required. It is the quality of this other provision by which most women in prison are likely to judge the service.

Psychiatric disorder in overseas women

Table 5.2 shows the diagnoses given to women from overseas, excluding three women for whom the interview was not completed because of language difficulties. No overseas women received a second diagnosis.

Thirty-one per cent of overseas women received a diagnosis, compared to 57 per cent of women resident in the UK. This

Table 5.2 Female sentenced prisoners who are ordinarily resident overseas: psychiatric diagnoses

Diagnosis	n	%
Neurotic disorders	10	28
Drug dependence/abuse	1	3
No diagnosis	25	69
Total	36	100

difference is statistically significant, the odds ratio being 3.0, with a 95 per cent confidence interval of 1.4 to 6.4. The numbers are rather small to allow statistical comparisons for individual diagnoses but the higher prevalence of drug dependence in women resident here reaches statistical significance (odds ratio = 12.3, with a wide 95 per cent confidence interval from 1.7 to 91). The difference in rates of neurotic disorder is not statistically significant.

The different pattern of psychiatric disorder in women resident in other countries validates the decision to analyse them separately and to exclude them from the comparison with male prisoners. None of the overseas women had a record of previous convictions.

Apart from a single case of drug dependence, the only diagnosis found in overseas women was neurotic disorder, usually depression or an adjustment reaction. The likelihood of developing depression is increased by adverse social circumstances and a lack of social supports – two characteristics of overseas women in prison, who often have language difficulties and rarely receive visitors. The bland description of adjustment reaction does not do justice to the profound distress which many of these women experience, often separated from children who face an uncertain future.

The low prevalence of drug dependence appears surprising at first sight, as most of these women were sentenced for the importation of heroin and cocaine. In fact, their involvement in drug smuggling stems almost entirely from economic motives. Most appeared naive about drugs and crime, and said their offence was a simple response to the offer of money for transporting goods about which they had asked few questions. Having been told by their employer that little risk was involved, and that they would be deported if caught, they were shocked by the reality of a lengthy prison sentence and it is not surprising that many became depressed.

Foreign drug couriers are not the central focus of the present report but they form a substantial minority within the female prison population and present a challenge to prison psychiatric services. The social isolation and language problems that contribute to their high rate of neurotic disorder may also act as obstacles to their receiving the counselling help and medical care that they need.

A comparison of psychiatric disorder in women and men

The following comparison is restricted to inmates ordinarily resident in the UK, for the reasons given in Chapter 3.

Prevalence rates

Five hundred and ten men (38 per cent) received at least one diagnosis, compared to 57 per cent of women. This difference is statistically significant. The female:male odds ratio is 3.2, the 95 per cent confidence interval being from 2.5 to 4.2. It is therefore possible to conclude that, when equivalent groups of female and male prisoners are compared, using identical methods, a much higher proportion of the women receive a psychiatric diagnosis. Of course this is not the full story. In order to understand the gender differences the figures must be broken down by diagnosis (Table 5.3).

Although the frequency of the various diagnoses may differ, the pattern is the same in men and women. The most common diagnoses are substance abuse, personality disorder and neurotic disorders, with psychosis accounting for only a small proportion of mental disorder. Note also that the prevalence of psychosis is approximately equal in the two samples, at just under 2 per cent. These points are illustrated in Figure 5.1, which presents the data as a graph.

Note that gender differences are found not in the rates of schizophrenia and other psychoses, but in neurosis, personality disorder, drug abuse and mental handicap. These four disorders are more common in women and the odds ratios suggest that these differences are statistically significant, i.e. they reflect true differences in the prison population. The difference between the samples is greatest for mental handicap, although the 95 per cent confidence interval for the population odds ratio is wide, with a minimum value of 1.4. This diagnosis is discussed in more detail below.

Table 5.3 The prevalence of psychiatric disorder in sentenced prisoners, by gender

	Women N = 258		Men N = 1751		Odds ratio	95% c.i.
	n	%	n	%		
Psychosis	4	1.6	34	1.9	0.8	0.3–2.3
Neurosis	40	16	104	6	2.9	2.0–4.3
Personality disorder	46	18	177	10	1.9	1.4–2.8
Alcohol abuse/dep.	24	9	203	12	0.8	0.5–1.2
Drug abuse/dep.	67	26	203	12	2.7	2.0–3.7
Mental handicap	6	2.3	11	0.6	3.8	1.4–10
Sexual disorders	1	0.4	38	2.2	0.2	–
Eating disorders	1	0.4	–	–	–	–
Epilepsy	1	0.4	4	0.2	–	–
Diagnosis uncertain	–	–	7	0.4	–	–

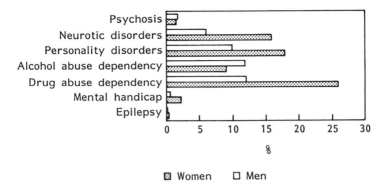

Figure 5.1 The prevalence of psychiatric disorder in sentenced prisoners, by gender

Odds ratios have not been calculated where the number of cases is very small. The single case of a woman diagnosed as suffering from a sexual disorder had started life as a man, but had undergone a sex change operation and now lived as a woman. The single case of eating disorder was a young woman suffering from bulimia nervosa.

Estimated number of cases in the prison population

So far, gender differences have been discussed in terms of prevalence rates or percentages, rather than the actual numbers involved. The figures in Table 5.3 can be used to calculate estimates of the total number of cases with a particular diagnosis, in the sentenced prison population as a whole. The study used a 1 in 4 sample of women and a 1 in 20 sample of men, so the figures are simply multiplied by 4 and 20 respectively. The results of this exercise are shown in Table 5.4.

These figures are rough estimates but they illustrate a crucial point. The difference in size of the two populations means that, even for diagnoses which are more common in women, the number of male cases is far greater than the corresponding number of women. This serves as a warning against drawing unwarranted conclusions about the operation of the criminal justice system from percentage figures, e.g. the high percentage of women in prison with a drug problem cannot be seen as evidence of the courts' greater willingness to imprison women with a drug problem, compared to similar men. This point is made even clearer when the figures are presented as a graph (Figure 5.2). The overall picture is quite different from that seen in Figure 5.1.

Table 5.4 Estimated total number of 'cases' in the sentenced prison population, by diagnosis and gender

Diagnostic group	Women	Men
Psychosis	16	680
Neurotic disorder	160	2080
Personality disorder	184	3540
Alcohol abuse/dependence	96	4060
Drug abuse/dependence	268	4060
Mental handicap	24	220
Other disorders	8	900
Epilepsy	4	80

These figures have important implications for service provision, and the level of demand facing the health service. The demand confronting the NHS is best represented by Figure 5.2, rather than Figure 5.1. If we consider only those inmates with a psychosis, it is obvious that the logistic difficulties involved in assessing and treating sixteen women with this diagnosis are of a different order of magnitude from the problem posed by 680 equivalent men. This point will be discussed in more detail in the following chapter. It is a problem which runs through any consideration of gender and service provision for mentally disordered offenders; women are always likely to be a minority, often a tiny minority, in terms of overall demand.

Before moving on to consider the implications of these findings for service provision, it is necessary to describe previous psychiatric treatment and the current treatment needs of the prisoners in the

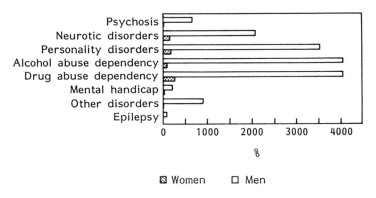

Figure 5.2 Estimated total number of 'cases' in the sentenced prison population, by diagnosis and gender

survey. Previous psychiatric treatment is described in two sections, treatment outside and inside prison.

Psychiatric treatment prior to imprisonment

All prisoners interviewed were asked if they had ever had any form of psychiatric treatment, or medical help for 'nervous trouble'. They were also asked if they had ever been in hospital for any reason, and full details of any positive responses were recorded. Whenever possible, details were then obtained from the hospital concerned.

Previous psychiatric contact was reported by 45 per cent of women, compared to 36 per cent of men (Table 5.5). Women were less likely to report no psychiatric contact and more likely to report previous outpatient or inpatient treatment as an adult. Men are more likely to report treatment only at a child guidance clinic, before the age of 16 years.

Table 5.5 Previous psychiatric treatment by gender and type of treatment

Type of treatment	Women		Men		Odds ratio	95% c.i.
	n	%	n	%		
None	143	55	1125	64	0.7	0.5–0.9
Child guidance only	22	9	291	17	0.5	0.3–0.7
Adult outpatient	61	24	199	11	2.4	1.8–3.3
Adult inpatient	32	12	136	8	1.7	1.1–2.5
Total	258	100	1751	100		

It soon became apparent that the commonest single reason for psychiatric contact was drug dependence and women had particularly high rates (see Table 5.2). In order to test the possibility that this difference alone could account for the gender differences in psychiatric treatment, the analysis was repeated after excluding all drug-dependent inmates (male and female). The results are shown in Table 5.6.

Even when all drug-dependent prisoners are excluded, women are more likely to report treatment at a child guidance clinic only. The difference in the proportion reporting no psychiatric contact is not statistically significant, as the gender differences in adult and childhood treatment balance each other out.

Table 5.6 Previous psychiatric treatment by gender and type of treatment, excluding all drug-dependent prisoners

Type of treatment	Women		Men		Odds ratio	95% c.i.
	n	%	n	%		
None	124	64	1062	68	0.8	0.6–1.1
Child guidance only	19	10	250	16	0.6	0.4–0.9
Adult outpatient	31	16	150	10	1.8	1.2–2.7
Adult inpatient	21	11	97	6	1.8	1.1–3.0
Total	195	100	1559	100		

There is nothing very surprising about these findings. Most of the child-guidance contacts were for conduct disorder, which is known to be more prevalent in male children. It would be unwise to place too great an emphasis on the precise figures as they rely on self-report and probably represent a considerable underestimate. Under-reporting is likely to be a particular problem with treatment in childhood, as prisoners were often being asked about events in the distant past.

The findings in respect of adult psychiatric treatment are broadly in line with gender differences for the utilization of psychiatric services by the population as a whole. For all mental illness in England and Wales, the rate of GP contacts for the age group 15–44 years in 1989 was 15 308 per 100 000 for women compared to 6298 per 100 000 for men, a ratio of 2.4 : 1. First hospital admission rates were 110.7 per 100 000 in women and 102.6 per 100 000 in men (1.1 : 1) (Department of Health, 1990).

Despite the methodological differences, it is interesting to compare the figures with those of Gibbens (1971). He found that 17 per cent of sentenced women had been psychiatric inpatients within the last three years and 8 per cent had been inpatients more than three years ago, i.e. a total of 25 per cent had been psychiatric inpatients, compared to the present estimate of only 12 per cent. The most likely explanation is Gibbens' use of a reception sample; mentally disordered women detected at reception may be transferred to hospital, and would not appear in a cross-sectional sample such as that used in the recent survey. However, the figures may also reflect changes in the use of inpatient psychiatric services, with a decreasing tendency to admit patients to hospital in the 25 years since Gibbens' survey.

Data on psychiatric treatment in the distant past are difficult to interpret because of the limitations of self-report and the use of rather crude measures. For example, all forms of inpatient and outpatient treatment are lumped together under these two headings, irrespective of the number of treatment contacts or the precise type of treatment. It is useful to consider more recent psychiatric treatment, during the current period of imprisonment, where it was possible to record more details.

Psychiatric treatment within prison

The issue of psychiatric treatment within prison is an emotive one, raising two conflicting concerns. On the one hand, there is concern that patients whose liberty is restricted may not be free to refuse medication. Critics accuse prison doctors of using medication excessively (Benn, 1983), to stifle dissent and maintain discipline, rather than to treat mental disorder (Glick and Neto, 1977). The opposing concern is that prisoners, who have restricted access to treatment and are not able to choose their doctor, may be deprived of the appropriate treatment for their clinical condition.

The legal position is clear. Medication within prison may only be given if the patient is freely consenting. The Mental Health Act, 1983 specifies the circumstances under which medication may be given to detained patients against their will, but none of these provisions apply within prisons. The Act applies to patients detained in hospital and no part of a prison constitutes a hospital for these purposes, even if it is commonly referred to as a 'prison hospital'. Medication may only be given against a prisoner's will under common law when it is necessary in an acute emergency.

These restrictions apply to all prisoners, even those who have been accepted for transfer to hospital under the Mental Health Act and are waiting for a bed to become available. Some critics view this as an unnecessary restriction which prolongs the suffering of mentally disordered prisoners, and have argued for an extension of the Act to allow the compulsory treatment of prisoners in certain circumstances. Others, myself included, believe that hospital care involves much more than the right to compulsory medication, and such a change would lead to even more difficulty in transferring patients out of prison.

The debate over medication within prison has generated a lot of heat but few data. Official figures (Prison Department, 1987) report only the total number of doses of medication dispensed within a prison. There is a consistent picture of higher prescribing rates per inmate within female prisons but the picture is incomplete. It will be argued here that the use of psychotropic medication in prison cannot be understood without considering both the prevalence of psychiatric disorder in prisoners and the use of non-pharmacological treatments. A full account of gender differences in this respect must also be placed in the context of psychiatric treatment outside prison.

All inmates in the study were asked whether they had ever received medical or psychiatric treatment within prison. Treatment with psychotropic medication during the current period of imprisonment was reported by sixty-eight women (26 per cent) and 131 men (8 per cent). This difference is statistically significant (the odds ratio = 4.4, with a 95 per cent confidence interval from 3.2 to 6.2). Table 5.7 breaks down these figures by type of treatment.

Table 5.7 Experience of psychiatric treatment during current prison sentence, by gender and type of treatment

Type of treatment	Women		Men		Odds ratio	95% c.i.
	n	%	n	%		
Medication – neuroleptics	16	6	57	3	2.0	1.1–3.5
Medication – antidepressants	41	16	54	3	6.2	4.0–9.6
Medication – anxiolytics	41	16	88	5	3.7	2.5–5.6
Any psychotropic medication	68	26	131	8	4.4	3.2–6.1
Psychological treatments	21	8	60	3	2.5	1.5–4.2

The figures show that women are more likely to have received treatment in prison with any one of three types of psychotropic medication. This finding is consistent with prison department statistics but provides more detailed information. One in four women have received psychotropic medication at least once during their current imprisonment.

Should this be a cause for concern? The answer is not straightforward but the study provides other, relevant information. First, the prevalence of psychiatric disorder is greater in female prisoners, suggesting that the demand for treatment will also be higher. Rather than passive recipients of treatment foisted upon them by doctors, women will often be seeking help. A description of attitudes towards treatment is contained in the next section.

Second, substance abuse is the commonest diagnosis, in both men and women. Even in those for whom it is not identified as a problem, consumption of alcohol, tobacco and illicit drugs is high among offenders. Many will be predisposed to seek chemical solutions to problems of psychological distress, so it is hardly surprising that, when the availability of other drugs is suddenly restricted, demand for prescribed remedies increases. The most common complaint made to us by women (and men) about prescribing was doctors' refusal to meet demands for sleeping tablets or other sedatives.

Third, apart from medication, women also had significantly higher rates of treatment by psychological methods, usually counselling or psychotherapy. It is difficult to imagine prisoners being coerced into psychotherapy against their will and this finding suggests a greater demand for treatment in women's prisons. It may be a source of criticism that only 8 per cent of women report receiving psychological help, about a third of the number given drugs. However, this is a criticism that can also be levelled at doctors in many other settings, including general practice, where it has been seen as quicker and cheaper to reach for a prescription pad than to provide 'talking treatments'.

Fourth, the gender differences within prison are broadly consistent with the utilization of psychiatric services outside prison, as reported in Tables 5.5 and 5.6. There is, however, a marked increased in the size of the gender difference, when comparing prisons to the outside world. 'Previous outpatient treatment' gives rise to a female:male odds ratio of 2.4 at most, whereas the differences for 'any drug treatment within prison' have an odds ratio of 4.4. Expressed in another way, 24 per cent of the women interviewed had a history of outpatient psychiatric treatment, and the figure for treatment with medication within prison was slightly higher at 26 per cent. By contrast, the number of men who received medication within prison (8 per cent) was lower than the 11 per cent who had received outpatient treatment in the past.

In other words, one would expect women to present a higher demand for psychiatric services within prison, as they have made greater use of services before arrest. The prevalence of diagnoses is consistent with this view. In addition, the study included a standardized assessment of neurotic symptoms, the Clinical Interview Schedule (CIS). Mean scores were 4.8 for men and 7.1 for women ($t = -5.26$, $p < 0.001$). The number of women scoring over 16 on this scale was thirty-five (14 per cent), compared to 103 men (6 per cent). There is a consistently higher level of psychiatric symptomatology among women in prison compared to men, mainly accounted for by anxiety and depression.

However, gender differences in psychiatric treatment within prison are not entirely due to differences in demand. The differences in treatment before arrest are less than the differences within prison. Imprisonment appears to accentuate the usual gender differences in use of psychiatric services. This is probably because medical services in women's prisons are more willing to respond to the demand for treatment. All hospital officers in women's prisons are trained nurses, which may encourage inmates to present emotional problems for treatment and aid their diagnosis. Also, the macho culture of male prisons, frowning on emotional expression, may act as a disincentive to seek treatment for anxiety and depression.

No evidence emerged of medication being given to inmates without their consent. It was not uncommon for interviewees to report that some other inmates were 'drugged up' or treated with large doses of medication but the interviewers did not encounter a single case of a woman reporting that she herself had been given medication against her will, despite asking this specific question of all inmates who reported taking medication at any time during their imprisonment. The same was found in the male sample.

The absence of forced medication does not mean that doctors working in prison should be complacent about the prescribing habits. There was wide variation in practice. Some doctors feared abuse of psychotropic drugs and were reluctant to prescribe any, others wrote frequent prescriptions. Fear of benzodiazepine dependence led some doctors to use phenothiazines or other neuroleptics to relieve anxiety. Women sometimes complained about inadequate explanations of the effects and side effects of medication. All are important points, which could be addressed by medical audit. These criticisms are not unique to doctors working within prisons and could also be made of some general practices.

Inmates' most frequent complaint in relation to medication was of alleged under-prescribing. Many women stated that the prison doctor refused to prescribe for anxiety, depression or insomnia. These symptoms were common, especially among inmates with a history of substance abuse. Objective evidence of under-prescribing is difficult to obtain. A crude comparison of treatment within prison and diagnosis does not reveal any striking discrepancy between the rates of a particular diagnosis and possible treatments. Perhaps the main concern should be that psychological treatments do not play a greater role in the management of common psychiatric complaints – both in and out of prison.

Attitudes to treatment and recommended treatment

All women who reported psychological problems were asked about their attitude towards treatment. Their answers were used to complete a rating scale, whose two highest points are: '3 – wants treatment, partially for side benefits' and '4 – wants treatment', Both ratings imply that the person concerned would attend voluntarily for suitable treatment if it were made available. The lower points on the scale are '0 – no desire for treatment', '1 – no desire but would accept if offered' and '2 – ambivalent about personal suitability for treatment'.

A rating of 3 or 4 was obtained by seventy-six women (29 per cent), compared to 273 men (16 per cent). This finding is statistically significant, the odds ratio being 2.3, with a 95 per cent confidence interval of 1.7 to 3.0.

There is a large subjective element to the rating scale but these findings are part of a consistent picture. Women in prison are more likely to report symptoms of distress, more likely to express a desire for treatment and more likely to receive treatment, compared to men in a similar situation.

The final part of the picture is formed by the researchers' treatment recommendation. This decision was made after completion of data collection in each case. The treatment options are described in Chapter 4.

Psychiatric intervention was recommended for 113 women (44 per cent) and 24 per cent of men (Table 5.8). All types of treatment were recommended more often for women but only the differences in 'treatment in prison' and 'further assessment' reached statistical significance.

Table 5.8 Treatment recommendations, by gender and type of treatment

Type of treatment	Women		Men		Odds ratio	95% c.i.
	n	%	n	%		
None	145	56	1334	76	0.4	0.3–0.5
Outpatient treatment	56	22	179	10	2.4	1.8–3.4
Therapeutic community	20	8	96	6	1.5	0.9–2.4
Transfer to hospital	12	5	52	3	1.6	0.8–3.0
Further assessment	25	9	90	5	2.0	1.2–3.2
Total	258	100	1751	100		

These figures are consistent with the earlier findings, in that women are seen as having a greater need for psychiatric treatment, mainly for minor problems such as depression and anxiety, which are commonly treated in the GP clinic or the outpatient department. There seems no reason to doubt the finding that these disorders are more prevalent in women serving a prison sentence. In the same way, the finding of no significant difference in the proportion of women and men who required transfer to hospital is consistent with the similar rates of serious mental disorder.

Ethnic origin and psychiatric diagnosis

When overseas residents were excluded, the study sample included thirty-three black women. Diagnoses such as psychosis and mental handicap are relatively uncommon, so any relationship between diagnosis and ethnic origin would have to be fairly gross in order to achieve statistical significance.

A psychiatric diagnosis was given to 42 per cent of black women and 59 per cent of white women in our sample. Table 5.9 shows the breakdown of diagnosis by ethnic origin, for white and black/Afro-Caribbean women. The number of women in other ethnic groups is too small for useful analysis.

Substance abuse is the only diagnosis in which the confidence interval for the odds ratio suggests that the finding is statistically significant. Black women in the sample have higher rates of psychosis and learning disability but the statistics suggest that this may be a chance finding.

If the larger sample of male prisoners is analysed in the same way, a similar trend emerges (Table 5.10).

Table 5.9 UK resident women serving a prison sentence: diagnosis by ethnic origin

	White N = 216		Black N = 33		White:Black odds ratio	95% c.i.
	n	%	n	%		
Psychosis	2	1	2	6	0.15	0.02–1.1
Learning disability	5	2	1	3	0.8	0.1–6.8
Neurosis	34	16	4	12	1.4	0.5–4.1
Personality disorder	40	18	6	18	1.0	–
Substance abuse	84	39	5	15	3.6	1.3–10
Any diagnosis	128	59	14	42	2.0	0.9–4.1

Table 5.10 UK resident men serving a prison sentence: diagnosis by ethnic origin

	White N = 1472		Black N = 182		White:Black odds ratio	95% c.i.
	n	%	n	%		
Psychosis	26	2	7	4	0.45	0.2–1.0
Neurosis	92	6	9	5	1.3	0.6–2.5
Personality disorder	167	11	9	5	2.5	1.3–5.0
Substance abuse	385	26	17	9	3.5	2.0–5.0
Any diagnosis	580	39	40	22	2.3	1.7–3.3

Again, black prisoners in the sample have a lower rate of most disorders but a higher rate of psychosis, 4 per cent compared to 2 per cent. As with the women, the 95 per cent confidence interval for the population odds ratio includes 1.0 at its extreme, so the finding may be due to chance.

If male and female prisoners are combined for a comparison of the prevalence of psychosis in white and black inmates, the rates are 1.7 per cent in white and 4.2 per cent in black inmates. The black:white odds ratio is 2.6, the 95 per cent confidence interval being 1.2 to 5.6. For the same comparison, $Chi^2 = 5.15$ (using Yates' correction) and $p < 0.04$.

Discussion

These findings may be due to nothing more than chance. However, the figures are consistent with the possibility that black sentenced prisoners, both female and male, have lower rates of most psychiatric disorders but a higher prevalence of psychosis.

In addition to the possibility that this is a chance finding, three explanations suggest themselves.

Misidentification of psychosis in black inmates
It has been suggested that psychiatrists may label the normal behaviour of some black people as psychosis. There is little evidence to support this explanation in relation to most studies of psychosis and ethnic origin, which have used standardized assessments of mental state. In the present study, diagnosis was heavily dependent on the psychiatrist's clinical interpretation, so this explanation cannot be discounted. The case histories do not suggest that any of the women given a diagnosis of psychosis had a short-lived or atypical disorder.

A higher rate of psychosis in the black population of England and Wales
Although controversial, there is an accumulating body of evidence to support this proposition in respect of UK residents of Afro-Caribbean descent (Wessely et al. 1991) and some studies suggest rates of schizophrenia may differ by a factor of fourteen (Harrison et al., 1988). This is not the place to document such evidence in detail. Psychosis in sentenced prisoners represents such a small proportion of psychosis in the community that, even if the population finding is valid, it is difficult to see it as a full explanation of the present results, without reference to the filtering process interposed between the general population and the sentenced prison population.

Differential access to psychiatric services
Informal and formal methods of diversion from the criminal justice system operate at various stages along the pathway from offence to prison sentence. The attitudes of both users and providers of services could operate at any of these levels to result in black mentally ill people being less likely to reach hospital than their white counterparts. Similarly, it has been suggested that the high number of black mentally ill people among those referred by the police under Section 136 of the Mental Health Act 1983 reflects a failure of other routes of access to mental health care (Fahy, 1989).

Apart from possible racism, contributory factors include variation in the presentation of mental illness (Littlewood and Lipsedge, 1981), cultural attitudes to mental illness (Hitch and Clegg, 1980), misdiagnosis by doctors in primary care (Burke, 1984) or an over-diagnosis by psychiatrists of cannabis psychosis in black patients (Thornicroft, 1990).

These three explanations are not mutually exclusive, and elements of each may be important. The present study provides no estimate of their relative importance, if any, and indicates only that there is a need for more information about the ethnic origin of sentenced female and male prisoners suffering from psychosis. Most psychotic inmates had been identified by prison doctors, and it would be a relatively straightforward task to monitor the ethnic origin of this small group.

Chapter summary

Among sentenced prisoners, a greater proportion of women can be given a diagnosis, but the pattern of mental disorder is roughly similar in both women and men. The common diagnoses are neurosis, personality disorder and substance abuse. Psychosis is found in less than 2 per cent of sentenced prisoners, whether male or female. These findings are consistent with other surveys of sentenced prisoners and this overall picture can be regarded as a well-established characteristic of the sentenced prison population.

The overall prevalence of disorders has been related to the treatment needs of the prison population. As psychiatric disorder is more common in female prisoners, it is not surprising that a greater percentage of the women have had psychiatric treatment in the past, have continuing treatment needs, and are receiving treatment in prison. However, gender differences in prison cannot be explained solely in terms of differences in demand. It appears that there is a greater willingness to provide treatment for women, on the part of the institution, and there may be a corresponding reluctance to provide treatment for men with similar problems. In the majority of cases, whether male or female, the treatment need is not for hospital admission but for treatments which would normally be provided by the general practitioner, outpatient clinic or substance abuse service.

The treatment needs of prisoners have been summarized in general terms but many questions remained unanswered. How do mentally disordered women get to prison? What is their experience of imprisonment and how does it compare with that of women who are not mentally disordered? How should NHS services be responding to women prisoners' unmet needs? The following chapters address some of these questions, and look at the cases of individual women in more detail.

6

Mentally disordered women in prison I: Psychosis and other severe mental disorders

Introduction

Planners need numbers. If the aim is to move psychiatric patients out of prison, then the figures in Chapter 5 are the best guide to the number of hospital beds required. However, statistics tell only a small part of the full story. Individual detail is lost when cases are combined to give percentages. This chapter, and the two which follow, are concerned with the meaning of the figures.

The approach used in this section assumes that research does not always need large samples, especially when it strays into the field of medical audit. On the contrary, when attempting to find out what has gone wrong in the operation of existing systems, it may be most useful to look closely at individual cases. The following account looks in more detail at the characteristics of mentally disordered women in prison, and asks critical questions about the links between prisons and psychiatric services in the world outside.

There are two major gaps in the picture given by the prevalence figures. The first relates to the reasons for mentally disordered people being in prison. How does a woman with schizophrenia find herself serving a sentence for theft? The question implies that something has gone wrong with the mechanisms that are supposed to keep psychiatric patients out of prison. Statistical analysis may provide clues, but it is only by examining the individual case that specific problems can be identified. (A useful analogy is with airline disasters. Plane crashes are rare, and a statistical analysis of all crashes is not the best approach to determining the cause. It is necessary to examine each accident in detail, to determine the particular causes, before looking to see whether similar factors apply to other cases.)

The second gap in the statistical picture concerns the personal experience of mentally disordered women in prison. It is easy to become complacent when dealing in numbers. A 2 per cent prevalence of psychosis does not sound like much, yet each case of a prisoner with schizophrenia may be a tragedy for the woman concerned. The case summaries in this chapter are used to make particular points, and to draw attention to the plight of the women involved.

It is obvious that many of the problems considered here depend on the type of psychiatric disorder. In order to examine these issues in detail, it is necessary to deal separately with different types of mental disorder. The present chapter is concerned with psychosis and mental handicap/learning difficulties. These two categories are considered together, as both diagnoses raise the possibility of a need for transfer to hospital. The following two chapters deal with personality disorder and neuroses (Chapter 7), and substance abuse (Chapter 8).

Psychosis

The representative sample included only four women with a firm diagnosis of schizophrenia, or other psychosis. There is no point in attempting further statistical analyses and it would be a mistake to draw far-reaching conclusions about the health or criminal justice systems based on such small numbers. Although these four women may not be typical of mentally disordered offenders at other points in the criminal justice system, it is reasonable to assume that they are typical of severely mentally ill women within the sentenced prison population. It will become clear that they are women for whom the existing systems failed. These four women should not have been in prison and each case can be treated as an inquiry into what went wrong. Viewed in this way, the small numbers are less important. Each case illustrates some of the problems arising within current provision for offender-patients.

It is necessary at this point to re-state an important principle of this study. No moral judgment is implied by the researchers' decision that these women should not have been in prison. There is no attempt to usurp the function of the court in sentencing them. At the time of interview, it was the researchers' opinion that they were medically unfit to remain in prison, because of psychiatric illness, in the same way that others are unfit for prison because of physical illness. After treatment in hospital, they could return to resume their sentence, if and when they were fit to do so.

The cases are described below. All four of these women were located in some form of special regime, two in a prison 'hospital', one in a disciplinary segregation unit and one in Durham H Wing. They could not survive in the ordinary prison environment, either because of their violent or disturbed behaviour, or because they were at risk of persecution by other prisoners. Severe mental disorder rarely passes unnoticed in the prison population, and the survey found that most inmates with schizophrenia, whether female or male, had already been identified by the prison authorities as mentally disordered. As the following accounts show, the fact that the cases had been identified does not mean that they were receiving appropriate help. Brief comments come at the end of each case, followed by a more general discussion of the issues raised by all four.

Case histories

Case no. 133
A 37-year-old woman who had served two weeks of a twelve-month sentence for the possession of cannabis with intent to supply. The police found her flat to be in a chaotic state, containing fifty grams of cannabis and a note from a friend asking for some of the drug.

She was born in Jamaica but had lived in London for at least twenty years. No details of her early life were available and she was reluctant to discuss this subject. Reports in her file showed that she was known to social services as a socially isolated woman with an eccentric lifestyle. Of her five children, two had been taken into care shortly after birth and the other three were with their estranged father.

Before this period of imprisonment, there had been three psychiatric admissions. On each occasion, her mental state was characterized by persecutory and hypochondriacal delusions, auditory hallucinations and prominent affective disturbance. Neuroleptic medication resulted in a partial recovery but she defaulted from outpatient treatment shortly after discharge.

While in prison on remand, she was floridly psychotic with persecutory delusions, auditory hallucinations, restlessness, anorexia and disturbed sleep. She responded to treatment with neuroleptic medication, which was accepted willingly. A psychiatric report, prepared by the NHS consultant from her catchment area after she had spent four months on remand, did not recommend a psychiatric disposal on the grounds that her psychosis had improved. The report emphasized her drug abuse and suggested that it was the cause of all her symptoms.

At interview, she was irritable and suspicious, asking repeatedly about the purpose of the study and about confidentiality. She revealed ideas of reference and believed that staff and other inmates were always laughing at her. She was reluctant to discuss her beliefs in detail and appeared moderately depressed. Despite continuing to take neuroleptic medication, the severity of her symptoms required that she was located in the prison hospital because of her excessive suspiciousness, hostility and quick temper. In a less supervised setting, she would accuse people of persecuting her and this had resulted in several violent confrontations. Hospital staff were very wary of her changeable moods.

The researchers gave her a diagnosis of schizophrenia, schizo-affective type (ref. no. 295.7 in ICD9).

Comment
1. The delay between arrest and preparation of the psychiatric report meant that her condition had improved by the time she was seen, so her catchment area psychiatrist was able to avoid recommending a psychiatric disposal. As a result, her district health service had escaped the cost of treating her acute psychotic episode, contrary to the general principle that mentally disordered offenders should be the responsibility of the health service.
2. Her present mental state left plenty of scope for psychiatric intervention; her symptoms prevented her being on normal location in prison and cannabis abuse and past failure to comply with treatment suggested likely problems on release. She would probably be released from prison with nothing more than an outpatient appointment. Hospital transfer while in prison would have facilitated proper psychiatric supervision in the future.
3. The diagnosis is complicated by cannabis abuse but the overall clinical picture suggests a chronic psychotic illness rather than a drug-induced psychosis. Cannabis may have exacerbated the psychosis but it is equally plausible that her psychotic symptoms had led to increased use of cannabis and bizarre behaviour had contributed to her offence being detected. Cannabis psychosis should be accepted as a diagnosis only when other possibilities have been excluded. Otherwise, psychosis will go untreated in many people who use this common drug.

Case no. 137
A 23-year-old woman who had served three months of a thirty-six month sentence for burglary and deception. Together with another woman, she had committed a series of burglaries, then used stolen credit cards. She had previous convictions for similar offences and saw her offending as 'a way of getting money. I couldn't survive on benefits'.

Her parents had separated soon after her birth and her brother suffered from schizophrenia. Raised by her grandfather, she was always a difficult child, jealous of her siblings, attention-seeking and prone to tantrums. Frequent truancy from secondary school and thefts from home and shops resulted in referral to child guidance clinic. She obtained one 'A' level.

She had previous psychiatric admissions at aged 21 and 22 years, the first following a serious suicide attempt. Despite depressive symptoms, schizophrenia was thought to be the most likely diagnosis and she was discharged on neuroleptic medication but failed to attend for follow-up.

There was no evidence to suggest that she was mentally disordered at the time of her offence or that mental illness played a direct part in her offending behaviour. Shortly after beginning her sentence, her behaviour became disturbed, resulting in several attacks on prison officers. She attempted to barricade herself in her cell and set fire to it. Thought disorder, auditory hallucinations and bizarre and persecutory delusions had been noted.

At interview, symptoms included profound depression, prominent suicidal ideation, retarded movements, poverty of speech, persecutory delusions and second and third person auditory hallucinations. She was contained in the prison hospital, accepting treatment with neuroleptic and antidepressant medication and awaiting transfer to hospital under Section 47 of the Mental Health Act 1983.

The researchers gave her a diagnosis of schizophrenia, schizo-affective type (ref. no. 295.7).

Comment
1. It is futile and absurd to ask the question 'Is X mad or bad?' The psychiatrist is qualified to comment on only half of this question, the other being more appropriate for a theologian or moral philosopher. To pose the question in either/or form is like asking whether a towel is red or wet: the answer may be both, either or neither. The English Mental Health Act 1983 *never* requires a doctor to take a decision about morality or responsibility, only to decide when mental disorder is of a nature or degree which requires treatment in hospital.

In the present case, mental disorder does not seem to have been a direct cause of offending. The stress of imprisonment may have contributed to a relapse of the pre-existing disorder. Whatever the causal links, the severity of symptoms and the high risk of suicide leave no realistic alternative but transfer to hospital, if the case is to be managed safely.

2. Following her relapse, and despite prompt referral by the prison doctor to the catchment area, this woman had spent at least a month in the prison 'hospital' awaiting assessment and transfer, representing a considerable suicide risk. The case is a clear illustration of delays in the NHS resulting in acutely ill patients being inappropriately detained in prison. It is ridiculous to blame the prison or its staff for this state of affairs.

Case no. 378
A 37-year-old woman who had served one month of a six-month sentence for shoplifting.

Few details of her early life were available but schizophrenia had developed in late adolescence, resulting in multiple hospital admissions. She had always been a difficult patient, often absconding from hospital and failing to comply with outpatient treatment, but there was no history of violence. Reports noted few delusions or hallucinations in recent years but prominent negative symptoms including poor self-care, social withdrawal and lack of motivation. One report described 'borderline subnormal intelligence' but she had in the past worked in sheltered employment as a typist.

She lived in a hostel for psychiatric patients and was reported to shoplift every day in a blatant manner, resulting in many convictions. The index offence occurred less than two months after release from prison following a similar offence. The psychiatric report prepared for the court by the NHS consultant responsible for her care mentioned her history of schizophrenia but stressed the current absence of delusions and hallucinations. Her prominent negative symptoms were described in the report but were not identified as features of schizophrenia. It was suggested that she needed prolonged care in a locked ward but she was not felt to warrant medium security and local facilities for long-term inpatient care did not exist. The report then went on to stress her non-compliance, her persistent petty theft and her responsibility for her actions, before stating that there was no psychiatric recommendation in this case.

At interview, she was located on a segregation unit and receiving treatment with depot neuroleptics. Prominent negative symptoms included poor self-care, flattened affect, poor social skills and poverty of speech: she repeatedly asked 'Am I fat? Do you think I am fat?' and made rather vague and uninformative replies to many questions. No delusions or hallucinations were apparent. She had no real insight into the fact that she suffered from a severe, chronic mental illness.

The researchers gave a diagnosis of residual schizophrenia (295.6).

Comment

1. This woman had been a psychiatric patient for many years and showed prominent, classical features of chronic schizophrenia. It can be argued that her persistent shoplifting is part of a general disregard for social obligations (common in schizophrenia) and this feature of her illness makes her unsuitable for care in the community. Reports stated that there was no long-term 'asylum' care available. The decision not to provide such care is an ideological one and ignores the patient's needs.
2. Her situation within prison was intolerable and indefensible. The unit on which she was located served a disciplinary function and had no nursing care. Some of the prison officers, who had no training in the care of the mentally disordered, openly made fun of her odd behaviour. There was a demoralized and cynical attitude among staff on the unit, who felt they were being asked to do an impossible and futile job, without proper training or support.
3. It is impossible to justify the imprisonment of a seriously mentally ill person for minor property offences, no matter how often the offences are repeated. There is no evidence that previous prison sentences had acted as a deterrent. The most likely outcome was that the same cycle would be repeated when she was released from prison. This is a blatant case of the psychiatric services 'dumping' on the prison system a patient who is not dangerous but does not fit easily into the existing pattern of community care.

Case no. 2099

A 38-year-old woman who had served two years of an eight-year sentence for arson with intent to endanger life. Having formed a dependent relationship with a woman who had helped her, she became very demanding. When the woman tried to reduce contact, she perceived this as rejection and threatened her, developing the delusion that the woman was her true mother. A series of threatening letters culminated in an arson attack on her acquaintance's home, when she was asleep upstairs.

The patient had been born into a family of travellers and brought up in a caravan. She never attended school and learned to read and write in prison.

Her first psychiatric admission was at the age of 19 years, as a result of a hospital order under the Mental Health Act, following a conviction for theft. She was said to have been 'depressed and of low intelligence'. Two subsequent admissions followed suicide attempts, and she was given a diagnosis of personality disorder. There was a history of alcohol abuse some years ago but no evidence of recent dependence.

Previous offences included a life-threatening assault on her 4-year-old daughter and an arson attack on her husband's home which was motivated by pathological jealousy. Her relationship with her husband and with several other people is said to have been stormy and characterized by violent outbursts.

At interview, she showed limited social skills and appeared of low intelligence but was literate (her IQ was 75). She stated that the offence had been motivated by her belief that the victim was her mother. She still suspected this to be true but was now less certain. She was moderately depressed but not suicidal and no other abnormal beliefs or hallucinations were apparent. She was located in a 'hospital' unit and was being treated with regular neuroleptic medication. Her medical records described her as highly vulnerable with rapid mood swings which made her a potential suicide risk.

The researchers gave her two diagnoses, paranoia (297.9) and personality disorder (301.9).

Comment
1. Psychiatric evidence did not appear to have been presented at her trial, when she pleaded not guilty, claiming not to have carried out the acts in question. It is an inevitable consequence of an adversarial system that the court will sometimes be unaware of facts relating to serious mental disorder at the time of sentencing.
2. As she did not disturb the running of the prison, she had not been referred for assessment by her local psychiatric services for possible transfer. This is a short-sighted approach. There is obvious mental disorder related to a serious offence which could easily have led to the victim's death. Public safety requires a full psychiatric assessment and adequate arrangements for follow-up and supervision before she can be released and there seems no point in delaying this process until release is imminent.
3. Cooperation with treatment probably acted to her disadvantage. If she did not comply with medication, her mental state would deteriorate and necessitate a full psychiatric assessment.

Discussion

The precise diagnosis in these complex cases may be debatable but there is no doubt that all four women suffered from serious and disabling psychiatric illnesses. They were voluntarily taking neuroleptic medication and one was on a waiting list for transfer to hospital. None was able to live in ordinary prison accommodation. In contrast, nineteen of the thirty-four men with a psychotic illness were held on normal location. However, as with the women, most of these men were recognized as abnormal by those around them. Psychiatric surveillance appeared to be more effective in the smaller female prison population. Women's prisons were generally more orderly and regulated. In the larger male prisons, with more anonymity, it was easier for severe mental illness to go undetected and untreated. Most of the male cases on normal location were found at overcrowded local prisons with a rapid turnover, where staff may have little personal knowledge of inmates and all services are over-stretched.

Schizophrenia, prisons and hospital closures
It has been suggested that the loss of beds in psychiatric hospitals in England and Wales since 1954 has resulted in many psychotic patients going to prison. The work of Penrose (1939), showing an inverse relationship across countries between the size of the mental hospital and prison populations, has been cited in support of this suggestion (Weller and Weller, 1988).

The notion that the populations of yesterday's asylums and today's prisons are somehow equivalent has always been rather dubious. For example, it ignores the fact that at least half of the patients in psychiatric hospitals were women, whereas the female: male ratio in prisons is about 1:30. Furthermore, the results of the present study suggest that the total number of sentenced female prisoners suffering from psychosis is only sixteen in the whole of England and Wales (the possible range, derived from the 95 per cent confidence interval, is from four to thirty-one). Given that more than 70 000 mental illness beds for people aged 18 to 64 were lost between 1954 and 1984, sentenced female prisoners suffering from psychosis are numerically insignificant in any explanation of where the chronically mentally ill are now located. It is estimated that there are 150 000 people suffering from schizophrenia in England and Wales (Bebbington and Hill, 1985). Assuming that half are women, then less than 1 in 2000 women suffering from schizophrenia (or 0.05 per cent) are serving a prison sentence at any one time.

The corresponding figure for men is 680 sentenced prisoners suffering from psychosis, with a possible range from 478 to 919. A similar rough calculation suggests that this is around 1 per cent of all men suffering from schizophrenia in England and Wales.

For both men and women, these proportions are approximate and are likely to be overestimates. The figure for psychotic prisoners is not restricted to those suffering from schizophrenia but includes all psychoses. Despite these inaccuracies, the figures make a useful point about the contribution of prison to the accommodation of the mentally ill. The sentenced prison population is almost insignificant in any account of where women with schizophrenia are located since the closure of the large mental hospitals.

The links between prisons and the psychiatric services
Low numbers should not obscure the tragic situation of those women in prison found to be suffering from serious mental illness. Two were recognized as suicide risks, one of whom remained in prison simply because of delays in achieving transfer (Case no. 137) and the other because her minimal interference with the prison regime was not felt to warrant referral for transfer (no. 2099). Case no. 133 can also be seen as a victim of delays in the NHS, in that her mental state improved on treatment, while awaiting assessment, to such an extent that hospital admission was no longer an urgent necessity. An additional factor in this case was cannabis abuse, put forward in the psychiatric report as a cause of her symptoms, despite evidence of chronic mental disorder. The diagnosis of cannabis psychosis contributed to the rejection of this woman by her catchment area consultant, yet the evidence for this diagnosis was slight. The Clunis inquiry was critical of doctors' readiness both to accept this diagnosis as an explanation of psychotic symptoms, and to wash their hands of the problem (Report of the inquiry, 1994).

These four cases present an unflattering picture of psychiatry and psychiatrists in general. Doctors are seen scrabbling to think of reasons for not treating patients and for not taking them into the beds which they control. The structural reasons for their attitudes will be discussed below, and they are often constrained by shortage of resources. Nevertheless, some responsibility must lie with individual clinicians. Questions of resources appear to be given priority without any thought for the needs of the individual patient or even the wider question of public safety. Each case contains at least one example of psychiatric practice falling short of the ideal.

The delay in transfer from prison to health service care has been recognized as a significant factor increasing the length of time which mentally disordered offenders spend in prison. A study of prison transfers to one special hospital showed that the mean time from medical recommendation to admission in 1960 was 1.3 months, whereas by 1983 it had risen to 7.6 months, an increase of 6.3 months (Grounds, 1991). A recent study of the remand population, described in more detail in the next chapter, also criticized the time patients had to wait before being transferred (Dell et al., 1993a,b; Robertson et al., 1993).

In understanding delays, or health services' refusal to accept patients, the characteristics of the women concerned are also important. Chronic illness, non-compliance with treatment, substance abuse and recidivism increase the chances of rejection by psychiatric services in male psychiatric patients (Cheadle and Ditchfield, 1982) and suggest a need for specialized provision. The problems of patients who are difficult to place have received increasing attention in the literature (Coid, 1991) without any noticeable easing of the situation.

The final chapter of this book returns to the question of service provision for difficult patients. First, the description will be broadened to include other psychiatric disorders. Psychosis accounts for only a small proportion of psychiatric disorder in prisoners and most offender-patients will have other diagnoses.

Mental handicap and learning difficulties

Diagnosis and terminology
This particular diagnosis is fraught with difficulty, as past diagnostic labels have often acquired pejorative connotations. Mental handicap was preferred, as it was the term used by the Royal College of Psychiatrists at the time of the study, and did not carry the implication of slowed normal development which is implied by the ICD9 label of mental retardation. The International Classification of Impairments, Disabilities and Handicaps (Wood, 1980) makes a distinction between disability and handicap, and the methods used in the prison study were more likely to detect handicaps rather than disabilities. The alternative term 'learning difficulties' is included, as many people prefer it, but it is considered too imprecise to be used alone. Many people have difficulty learning some things, especially in a prison population where a history of schooling disrupted by social or personal problems is the norm.

The diagnosis of mental handicap in the present study was reserved for cases where there was evidence of significant handicap. Test results were taken into account when available, but low IQ alone was not considered sufficient to make the diagnosis, which, like the others, was made on the basis of clinical criteria derived from the International Classification of Diseases (WHO, 1978). As was also the case for psychosis, the small numbers preclude statistical analysis. The best way to illustrate the way in which the diagnosis was used is by considering the six women who were given this label, and they are described below. The possibility of additional psychiatric diagnoses is a major concern for the psychiatrist faced with a mentally handicapped patient (Corbett, 1979), so it is mentioned in the comment following each case.

Case histories

Case no. 382
A 29-year-old woman who had served two months of a twelve-month sentence for arson. Under various social pressures, including sexual advances from her brother, she set a fire in her own flat, then informed her neighbour. She had a previous conviction for arson in similar circumstances.

She had been in the care of the local authority from 2 to 11 years of age and attended special schools (her IQ was 58). As a child, she was a victim of long-term sexual abuse by her brother and father. She had three children, all adopted away shortly after delivery. She had several previous psychiatric admissions, usually following overdoses precipitated by social stresses, and had cut herself on many occasions both in and out of prison. She was a long-term attender at a mental handicap clinic and a social services day centre. One psychiatric report mentioned 'voices telling her to set fires' but found 'no other evidence of psychosis'.

On remand, she set fire to her own mattress and cut herself repeatedly. A medical report from a prison doctor described her as 'mentally impaired' with 'intellectual functioning below normal' but made no medical recommendation on the grounds that the patient 'would quickly arrange for a Mental Health Review Tribunal' and 'no Tribunal would agree to keep her in hospital'.

At interview, she was slow to respond and had poor social skills. Her reading age was seven years, she reported no delusions or hallucinations and was not on regular medication.

The researchers gave her a diagnosis of mild mental retardation (ref. no. 317 in ICD9).

Comment

1. There was insufficient evidence to justify a diagnosis of psychosis or other psychiatric disorders as the arson and self-destructive behaviour could be explained as responses to social stresses in a person of limited intellectual and coping ability. The research panel agreed with the prison medical officer that the patient was mentally impaired, as defined in the Mental Health Act 1983, but believed that hospital transfer would be the best treatment, because of the risk of suicide within prison and her need for long-term treatment and supervision.

2. The comments in the prison doctor's report are inappropriate. Quite apart from possible inaccuracy (it is unlikely that a Tribunal would allow this woman her freedom against medical advice), they represent an attempt to usurp the function of the Tribunal. Medical evidence to a court should be confined to matters in which the doctor has expertise, i.e. the diagnosis and treatment of mental disorder. The effect of the doctor's comments was to prevent transfer to hospital, depriving this woman of the treatment she needed.

Case no. 452

A 23-year-old woman who had served one month of a five-month sentence for shoplifting. She had several previous convictions and one previous sentence, all for theft. Born into a large, deprived and criminal family, she attended special schools but had no psychiatric history. She had never worked and remained dependent on her mother, who was the main care-giver to her 18-month-old baby. She coped well on normal location in prison, with the help and protection of her older sisters who were also serving sentences. She had been disciplined for intimidating other inmates, along with her sisters.

At interview, her speech was slow and inarticulate and she was unable to read any words on the Schonnell test.

The researchers gave her a diagnosis of mild mental retardation (317).

Comment

1. There was insufficient evidence to justify an additional diagnosis of personality disorder as her violent behaviour could be seen as resulting from social or other situational factors. She coped well in prison and there seemed to be little scope for medical intervention, although there is an obvious role for educational services.

Case no. 494
A 23-year-old woman who had served eighteen months of a thirty-month sentence for aggravated burglary. In the company of another woman, she had entered the home of a 70-year-old man, tied him up and assaulted him in the course of a robbery. Her accomplice had worked for the man and was the instigator of the offence. She had previous minor theft convictions but this was her first time in prison. She was from a chaotic family and had attended a special school. She had no history of psychiatric treatment but reported one overdose as a child and many episodes of self-cutting, mainly within prison. While in prison, she was exploited and bullied by other women and was therefore held on a segregation unit for her own protection.

At interview she was over-familiar and socially inappropriate in manner. She expressed no remorse for the index offence and minimized its seriousness but was angry at the length of her sentence. She described hitting out at anyone who annoyed her within prison and was preoccupied with lesbian approaches from other women, to the extent that this was an over-valued idea, although she showed no evidence of delusions or hallucinations. Her reading age was six years.

The researchers gave her diagnoses of mild mental retardation (317) and personality disorder (301.7).

Comment
  1. In this case, the extent of the violence, self-harm and relationship difficulties was thought to justify the additional diagnosis of personality disorder.

Case no. 496
A 44-year-old woman who had served fourteen months of a four-year sentence for offences of indecent assault against her son (aged 8 years) and daughter (15 years). This was her first conviction; she was found guilty of being a willing partner in her husband's more serious sexual offences against the children. She had attended special school and was known to mental handicap services, working at an adult training centre. Her husband also had learning difficulties and both received support from local services for the mentally handicapped and from social services. She had no psychiatric history. Reports at trial noted low intelligence but concluded that hospital treatment was not required and recommended probation. She had been bullied in prison (because of her obvious vulnerability, in addition to the nature of her offence) and was segregated for her own protection.

At interview, she appeared slow, nervous and ineffectual. Her reading age was six years and four months. She denied any role in her husband's offences, describing him as violent and forceful: 'I didn't know what he was up to but I couldn't have stopped him anyway.' She was divorcing her husband. She had no major problems coping with the prison regime on the segregation unit but felt she could not survive on normal location.

The researchers gave her a diagnosis of mild mental retardation (317).

Comment
1. There was no evidence for an additional diagnosis. Her husband appeared to have been the prime mover in the offences but the court had decided that she also bore considerable responsibility for what had happened.
2. The lengthy prison sentence may serve the purposes of retribution and deterrence but it leaves many other issues unresolved. There is clear evidence of mental disorder and there may be scope for therapeutic intervention aimed at reinforcing appropriate sexual boundaries and increasing her ability to assert herself with male partners. In an ideal world, treatment would be carried out within prison. In reality, it may require transfer to a hospital with expertise in this area.

Case no. 497
A 23-year-old woman who had served five months of a nine-month sentence for arson. She had many previous convictions and had served two previous sentences for criminal damage and assault.

She attended special school and showed difficult and aggressive behaviour from an early age. She had many psychiatric admissions, often compulsory, for episodes of bizarre and violent behaviour and was given provisional diagnoses of personality disorder, mental handicap, depression and, on one occasion, schizophrenia. Her IQ was measured at 66. She suffered from mild diabetes but was unable to conform to dietary restrictions. Repeated assaults on staff and her low intelligence led to rejection by local hospitals and hostels. Local psychiatrists agreed treatment was necessary and suggested referral to special hospital. She had carried out many assaults on prison staff, some of them serious.

This woman was interviewed in a special disciplinary unit, where she had spent the whole of her sentence. She had poor social skills and showed mild disinhibition but no delusions or hallucinations. She accepted neuroleptic medication, which was prescribed to reduce her aggression; she reported that it made her calmer and helped control her temper. Her reading age was six years. The researchers gave her two diagnoses, mild mental retardation (317) and personality disorder (301.9).

Comment
1. Despite obtaining extensive reports on her psychiatric history, there was insufficient evidence for a firm diagnosis of psychosis, though it remains a possibility. Reports suggested she needed long-term care at a level of security intermediate between the locked ward of a local hospital and the maximum security of a special hospital. This case exposes a local gap in provision between low and maximum security; a placement in a medium secure unit would have been suitable.
2. A special hospital opinion was not obtained and no efforts were being made to secure her transfer, as the prison medical officer did not believe that this was warranted. He had opposed earlier attempts to secure a hospital placement, on the grounds that she would escape her just deserts. This punitive attitude is indefensible and represents a further example of doctors commenting on matters outside their field of expertise.
3. A further problem may be that the case fell somewhere between the mental illness and mental handicap services. Each had tried to suggest that responsibility lay with the other. There are few incentives for services to seek patients from prison, a problem that will be addressed in the next chapter.

Case no. 499
A 29-year-old woman who had served two-and-a-half years of a five-year sentence for arson, having set fire to the hostel where she was living. She had served five previous custodial sentences, all short, for offences of arson (1), criminal damage (3), burglary (1) and theft (1).

After being abandoned by her own mother at the age of 9 months, her several foster placements had proved unsatisfactory, largely because of her behavioural problems. She attended special schools, where her behaviour was also difficult. Her IQ was assessed at 73 when she was aged 7 years. At aged 16 to 18 years, she lived in a hostel for the mentally handicapped but had difficulty forming relationships.

Brief psychiatric admissions dated from the age of 17 years, with numerous episodes of self-harm by cutting of her arms and attempted hanging, both in and out of custody. She showed aggression towards property (originally breaking windows, then progressing to arson) and people (kicking and punching, often directed towards hostel or prison staff). Possible short psychotic episodes, with suspected auditory hallucinations, responded rapidly to medication. Difficult behaviour resulted in rejection by local psychiatric facilities. Admission to a regional secure unit was felt to be inappropriate because of her low intelligence and probable need for long-term care. During the present sentence, many episodes of violence and self-harm resulted in periods of solitary confinement. At interview, she was placed in a special disciplinary unit within the prison. She had very limited social skills and a childlike manner. Symptoms included depression and anxiety, accompanied by frequent (at least daily) impulses to injure or kill herself and difficulty in controlling her temper when frustrated. She was unable to attempt the Schonnell reading test. No delusions or hallucinations were apparent but she was being treated with long-term neuroleptic medication which she said helped her keep calm.

Her release date was approaching and she had been referred to a special hospital and accepted for transfer, under Section 47 of the Mental Health Act 1983, on the grounds of psychopathic disorder.

The researchers gave her diagnoses of mild mental retardation (317) and personality disorder (301.9).

Comment
1. Despite extensive medical records from previous admissions, the evidence was not felt to justify a current diagnosis of schizophrenia or other psychosis. Follow-up information was obtained later, after her move to hospital. The conclusion was that schizophrenia was a likely additional diagnosis. One year after transfer, she killed herself within a special hospital.
2. There was no justification for hospital transfer to be delayed until the end of her sentence. Again, low intelligence and the chronic nature of her problems appear to have made her unattractive to psychiatric services. The practice of hospital referral as a prisoner's date of release approaches has been criticized elsewhere and is impossible to justify in the present case, where the woman's behaviour appeared to have been worse during the early part of her imprisonment.

Discussion

All these cases show evidence of low intellectual ability, resulting in a significant social handicap. As was the case for psychotic inmates, the mentally handicapped women detected in the survey did not blend in with the 'normal' prison population. They required some form of special management and a high proportion were also suffering from psychiatric disorders that required treatment.

Three of the cases (nos 382, 497 and 499) received diagnoses of schizophrenia or another psychosis when assessed elsewhere, either before or after assessment in the study, and two were receiving neuroleptic medication. It is apparent from the case histories that in many respects they were more disturbed and socially disabled than some women who did receive a diagnosis of psychosis.

This degree of diagnostic uncertainty is relevant to the findings on psychosis. If all three of these women had been given a diagnosis of a psychotic disorder in addition to mental handicap, the total number of psychotic women would have increased to seven (2.7 per cent of the sample). This would not be a significant difference from the 1.9 per cent of men with this diagnosis, even before allowance is made for the fact that some of the men with a diagnosis of mental handicap may also warrant an additional diagnosis.

At least three of these women would have been more appropriately placed in a psychiatric hospital. As noted above, one was transferred to special hospital after our study. The small number of cases should warn against looking for general principles to explain their failure to reach hospital. However, two of the cases illustrate the point made by Allen (1987), that prison medical officers are sometimes reluctant to recommend psychiatric treatment, even when severe disorder is apparent. Allen refers to reports that end with 'the implicitly damning conclusion: "This offender is fit for any penalty that the court may impose".' This phrase appeared in the reports prepared by medical officers on cases 382 and 497, where it serves also as a damning reflection on the standard of some psychiatric reports. Allen believed that this type of report was generally confined to male offenders; the present study confirms that it is not.

All six women with a diagnosis of mental handicap were held in one prison and had passed through a single remand centre (the eleven men with a similar diagnosis were contained in seven different prisons). The three most disturbed women were held on the same unit, for inmates who present control problems, and two of them were the responsibility of the same health region. As the female prison population is so small, the policies of a single prison, health region or even an individual doctor can exert a disproportionate influence on the number of mentally disordered inmates in the prison system. It is possible that the policies of doctors at one prison contributed to the high proportion of mentally handicapped inmates found in sentenced female prisoners. The evidence for this proposition is to be found in the reports expressing active opposition to proposed hospital referral and transfer. This was a very unusual occurrence, most prison doctors being glad to be relieved of the responsibility of looking after mentally disordered inmates.

A study of male remand prisoners (Coid, 1988) found considerable variation in the frequency of hospital disposal both between health regions and between individual psychiatrists. The same study also found that patients with a learning disability were particularly likely to be rejected by their catchment area psychiatrist, because of a perceived need for long-term care.

If the figures in the current study are accepted at face value, as representative of the female prison population, they suggest that the total number of mentally handicapped women serving a prison sentence is approximately twenty-four, with a wide 95 per cent confidence interval from 8 to 42. The equivalent figure for the sentenced male population is 220 (95 per cent c.i. from 73 to 367). These numbers are very small compared to the number of mentally handicapped people in the community, or the number of beds lost in mental handicap hospital closures over the last thirty years.

Robertson (1982) describes a decline in the use of the criminal provisions of the Mental Health Act in relation to offenders with learning difficulties since the 1960s. Mental handicap hospitals have concentrated their resources on the care of severely handicapped patients and are less likely to admit the mildly handicapped. Robertson suggests that most offenders with learning difficulties have not been disadvantaged by this change; in the past, it is likely that many were hospitalized unnecessarily. This may be true in most cases. However, the present findings suggest that these changes have left a gap in provision, for the mildly mentally handicapped offender with a behavioural or psychiatric disorder. These cases fall uncomfortably between mental handicap and mental illness services. Facilities for learning disability patients were unable to cope with a high level of disturbed behaviour, whereas mental illness hospitals were unwilling to offer long-term inpatient care. In this way, a learning disability acts to increase the chances of a mentally disordered offender being rejected by psychiatric services.

Chapter summary

This chapter has been concerned with psychiatric disorders that are uncommon in prisoners, but are very likely to require transfer to hospital for safe and effective treatment. Prisoners with serious mental disorder are highly visible within prison, but the reason for their presence is often to be found in the health service, which makes insufficient provision for difficult patients with chronic disorders. These points will be taken up again in the final chapter. The following two chapters deal with more common psychiatric disorders found in prisons.

7

Mentally disordered women in prison II: Personality disorders and neurotic disorders

Introduction

Both personality disorder and neurotic disorders are common in prisoners, and most cases are treatable within prison. This distinction is not absolute, as more severe cases of personality disorder and, occasionally, neurotic disorder will require transfer to hospital, in which case they encounter the problems described in the previous chapter, and similar comments apply. In most cases, though, the focus is on the way in which these disorders are handled by services within the prison.

The combining of personality disorder and the neuroses may seem arbitrary but can be justified by the fact that they have many symptoms in common. Anxiety, depression, low self-esteem and the impulse to self-harm may characterize both conditions, and present a challenge to medical services within prisons.

Personality disorder

Personality disorder is a more controversial diagnosis than psychosis or mental handicap. It is difficult to establish acceptable levels of reliability and validity. Some psychiatrists use the term in a pejorative sense and it has sometimes been regarded as synonymous with criminality. In the circumstances, it is tempting to abandon this diagnostic label altogether.

Unfortunately, the concept of personality disorder is not easily discarded. After substance abuse, it was the most common diagnosis given to women in prison. It was the only diagnosis given to eighteen women (7 per cent) and it was given as an additional diagnosis to twenty-eight others (11 per cent). The equivalent figures for men are 104 (6 per cent) with this diagnosis alone and seventy-three (4 per cent) given it in addition to other diagnoses. The gender difference is statistically significant (see Chapter 5). No account of psychiatry in relation to female prisoners can be complete if it does not deal with personality disorder.

The problems raised by the diagnosis are of two types. First, there are inherent contradictions in the concept of personality disorder and its relationship to criminal behaviour. Any study, whatever its methods, runs into these difficulties. Second, there are the problems raised by the specific methods used in the prison survey, i.e. the use of clinical criteria to identify personality disorder. These problems will be considered in greater detail below, after three examples of the way in which the diagnosis was used.

Case histories

Case no. 95

A 24-year-old woman who had served four months of a fifteen-month sentence for theft and robbery. She grew up in children's homes, her first conviction at 13 years being followed by many convictions for petty theft and three for violence, resulting in two previous prison sentences. She began to drink alcohol at the age of 12 years and reported drinking 'until I'm paralytic, if I've got the money'. She described symptoms of withdrawal when she entered prison, requiring treatment in the prison hospital. She was a regular user of heroin ('every day') and cocaine, amphetamines and benzodiazepines ('most days'), and saw her drug use as a major problem. She had been in a drug treatment centre but could not tolerate the regime and left after a few days. Her index offence was a street robbery of another woman, accompanied by gratuitous violence (kicking and punching): 'I'm sorry now but I don't really remember it, I was so out of my head on drink and drugs. That's where the money went.'

At interview, she was softly spoken and intelligent, revealing long-standing feelings of depression, tension and irritability and a history of overdosing on numerous occasions. She described herself as having no friends, frequently arguing with staff and other inmates, often thinking that other people were looking at her or trying to annoy her.

The researchers gave her a total of three diagnoses: drug dependence (ref. no. 304.7 in ICD9), alcohol dependence (303) and personality disorder (301.9).

Comment
1. The history of relationship difficulties, violence and persistent psychiatric symptoms were judged to warrant a diagnosis of personality disorder in addition to the substance abuse diagnoses.
2. Various psychiatric symptoms have been present for many years and cause problems both for this woman and people with whom she comes into contact. The symptoms cannot easily be dismissed, whatever diagnostic problems they raise.

Case no. 143
A 26-year-old woman who had served nine months of a forty-two month sentence for robbery, having entered a flat and stolen a ring from the female occupant before hitting her over the head with a glass ashtray. Her parents were unknown and her early life had been spent in children's homes, foster placements breaking down because of aggressive behaviour which resulted in two referrals to a child guidance clinic. She absconded repeatedly from residential schools, becoming involved in prostitution and drug use. She had many convictions from age 15 for theft, deception, burglary and assault, often accompanied by heavy drinking which she said was 'no longer a problem'. She described herself as having been 'obsessed with shoplifting, just for its own sake, taking things to throw away'.

At interview, she was located on the mother and baby unit with her seven-month-old daughter, born during the present sentence. Two previous children had been adopted away and a third was permanently in the care of the father as 'I don't want to see him or it'. Her manner was aggressive and confrontational and she complained of many physical problems and of her temper and irritability, which caused her problems 'every day', both in and out of prison. She reported other people making her angry 'all the time. It's been a problem all my life, though it's not as bad right now, in here'. She attributed her previous convictions for assault to her loss of control, which meant that 'rows get out of hand', and to other people provoking her. Previous relationships had ended because of her outbursts. She accepted no responsibility for the index offence, claiming that she had been 'forced to do it', by financial necessity. She believed the victim had provoked the violence by resisting the theft.

The researchers gave a diagnosis of personality disorder (ref. no. 301.7 in ICD9). She denied that alcohol abuse was a current problem and there was insufficient evidence to support this additional diagnosis.

Comment
1. As in the previous example, her symptoms caused problems to herself and to others. Violence is more common in male prisoners with a personality disorder but it cannot be assumed that women will not present significant problems of this type.

Case no. 117
A 19-year-old woman who had served one month of a six-month sentence for burglary and assault. Brought up in a middle-class family, she was described as 'wild and uncontrollable' from childhood, requiring special schooling. Her convictions began at the age of 16 years and included five thefts and three assaults, resulting in four previous prison sentences, all marked by repeated offences against discipline, including assaults on staff.

At interview, her manner was loud and she was verbally aggressive. She described long-standing problems with her temper and outbursts of violence 'for no reason. People just say something and it winds me up'. She described difficult relationships with staff and inmates, saying that this was partly her temper but also their own fault 'for provoking me'. She was not currently depressed but described past episodes of depression and tension when she would cut her arms repeatedly, both inside and outside prison.

The researchers gave her a diagnosis of personality disorder (reference no. 301.7 in ICD9).

Comment
1. Again, there is significant violence towards others, motivated largely by internal events. Self-harm and difficulty relating to others were the main additional reasons for making the diagnosis. Her behaviour was seen as inappropriate by both the interviewer and the woman herself.

Discussion

The relationship between personality disorder and crime
The confusion between personality disorder and criminality has a long history. Maudsley (1874) described psychopathic disorder as a mental disorder with 'so much the look of vice or crime that many persons regard it as an unfounded medical invention'. Criminal acts are regarded as evidence of personality disorder in certain circumstances. This problem is not easily resolved and it may be that it is, in principle, insoluble.

It is not surprising that studies of prisoners run into difficulties as a result of this confusion. Previous surveys of male prisoners report rates of 'psychopathy' or 'sociopathy' ranging from 5 per cent (Jones, 1976) through 8 per cent (Roper, 1950) to 78 per cent (Guze, 1976). The results vary according to the definition used and, in particular, the extent to which criminality is used as an indicator of antisocial personality disorder. The same problem is encountered in studies outside prisons and is not solved by using standardized instruments. The prevalence rates for antisocial personality disorder in the ECA study (Regier et al., 1988), 0.2 per cent for women and 0.8 per cent for men, are partly a restatement of the observation that women offend less frequently than men.

In a prison survey concerned with the need for psychiatric services, it is pointless to restate the fact that women have lower levels of offending than men. Therefore, when making the diagnosis of personality disorder based on clinical criteria, the researchers attempted to disregard criminal activity. Greater weight was given to the presence of psychiatric symptoms. In the case histories above, it can be seen that the diagnosis depends on the presence of symptoms, not offending.

Of course, the use of clinical criteria in this way, rather than one of the standardized instruments for assessing personality disorder, reduces the reliability of the findings. They remain the findings of a particular research team, and others may have taken a different view. The positive aspect of this method is that the findings have good face validity. They represent the views of mental health professionals, applying the standards they use in their normal practice.

Personality disorder and psychiatric services
The rationale for a clinical approach to diagnosis is its relevance to psychiatric practice. This can be tested by asking how many of the women given the diagnosis have experienced psychiatric treatment in the past. Current attitudes to treatment also serve as a guide to the extent to which the women themselves believe that they have problems that may appropriately be presented to a psychiatrist.

Of the forty-six women given the diagnosis, thirty (65 per cent) had a history of psychiatric contact before entering prison, including nine of the eighteen in whom personality disorder was the only diagnosis. In the same way, most of the women given this diagnosis believed that they needed some form of psychiatric help. Although eight women (17 per cent) were judged to have no desire for treatment at present, twenty-two (48 per cent) were judged to have a strong desire for treatment, with little interest in side benefits. The other women with the diagnosis had an intermediate level of motivation, but stated that they would accept treatment voluntarily, if it was offered.

The important point is that these findings cannot be dismissed as 'psychiatrization' of offenders by doctors. Most women given the diagnosis reported that they would accept psychiatric treatment if it were offered, including many who expressed a strong desire for treatment. These women may have resented the label and may not have construed their problems in the same terms as a psychiatrist but their willingness to accept medical intervention suggests a degree of common ground. Over half of these women had already received psychiatric treatment outside prison. Any comprehensive psychiatric service to prisoners would be required to respond to the problems these women presented. The researchers' recommendations for psychiatric treatment are shown in Table 7.1, which compares women and men with the diagnosis.

The eight women with personality disorder who were recommended for hospital transfer included three with an additional diagnosis of mental handicap (see preceding section). None of the other women (or men) with a diagnosis of personality disorder had additional diagnoses other than substance abuse or neurosis.

Most women with personality disorder were recommended for outpatient treatment, which could include various forms of psychological treatment, counselling and psychotherapy. The initial stages of further assessment could also be done as an outpatient. Only two women were judged to have no treatment needs.

Table 7.1 Prisoners given a diagnosis of personality disorder: treatment recommendations by gender

	Women N = 258		Men N = 1751	
	n	%	n	%
'Outpatient' treatment within prison	15	6	44	2
Therapeutic community	9	3	46	3
Further assessment	12	5	35	2
Transfer to hospital	8	3	9	< 1
None	2	1	43	2
Total	46	18	177	10

Comparison with other studies

A high rate of personality disorder in female prisoners is in accord with the findings of Gibbens' (1971) study, which refers to many women suffering from 'personality or behaviour disorders'. Dobash, Dobash and Gutteridge (1986), in a critique of his study, argue that the high prevalence of personality disorder was one of the reasons for the failure of the Holloway project. When Gibbens spoke of a high rate of psychiatric disorder, he was actually referring to a high rate of personality disorder rather than other diagnoses. The response of psychiatry to these patients is ambivalent at best and another study in which Gibbens was involved found that only 9 per cent of women remanded to Holloway for medical reports were accepted for treatment, while 53 per cent of those identified as suffering from personality disorder were rejected. A psychiatric approach to these patients was unlikely to prove successful when psychiatry did not have clear and generally accepted ways of approaching their treatment.

Carlen (1985) criticizes the concept of personality disorder on the grounds that it attributes disturbed behaviour to the individual. An alternative is to see the behaviour as a reaction to prison life, which can best be explained by analysing the prison regimes which 'are enough to drive any woman crazy'. The present study cannot answer questions about the cause of psychiatric disorder. However, Carlen's suggestion cannot explain the findings, as the diagnosis of personality disorder was not given to women whose disturbed behaviour was confined to prison. It was unusual for behaviour such as deliberate self-harm to be manifest only within prison and the case histories described above reveal chronic symptoms that cannot be explained away as a response to imprisonment.

In a more cogent criticism, Carlen (1985: 619) also draws attention to the contradictions inherent in psychiatry's use of the concept of personality disorder, the 'simultaneous identification of psychopathic personality disorder and its refusal to recognize it as a category of treatable mental illness'. An important finding of the present project, whatever the problems of diagnosis, is that most of the women so labelled were judged to require some form of psychiatric help, ranging from hospital transfer to counselling (Table 7.1). Much of this help could be provided, and is provided, by non-medical personnel. The contradiction identified by Carlen would be largely resolved if psychiatrists were to acknowledge the treatment needs of these women and abandon the 'therapeutic nihilism' (Lewis, 1974) which has often characterized medical attitudes to this diagnosis.

Neurotic disorders

A diagnosis of neurotic disorder was made in forty women (16 per cent) and 104 men (6 per cent). Neurosis was the only diagnosis in twenty-four women (9 per cent) and fifty-seven men (3 per cent); it was given as an additional diagnosis to sixteen women (6 per cent) and forty-seven men (3 per cent).

As with personality disorder, interviewer bias may have played a part in producing these results. An attempt at validating the findings can be made by comparing the scores on the Clinical Interview Schedule (CIS) (Goldberg et al., 1970), which is a good measure of the breadth and severity of neurotic symptoms. The researchers had the CIS score available to them when making their diagnosis, although the final decision was made on clinical criteria. It would be evidence of bias if women were shown to be given the diagnosis at a lower score (i.e. level of symptoms) than men.

In fact, the mean score for women given a diagnosis of neurotic disorder was 17.2, as against 17.9 for men, i.e. there was no significant difference. However, there was a gender difference in the scores of those who were not given the diagnosis: the mean for women was 5.3, against 3.9 for men. The 95 per cent confidence interval for the female:male difference in the scores of 'non-cases' is 1.1 to 1.7, suggesting that this is not a chance finding.

The interviewers' impression was that men were more likely to respond to the interview by denying *all* symptoms of depression or anxiety. In the stressful environment of prison, this may be seen as a primitive defence against anxiety, which was used more often by men. An additional factor was the use of two male interviewers throughout the study. Men in prison may have felt obliged to maintain a facade of coping, when questioned by another man, in a way which women did not.

This is not the place in which to discuss in detail the use of standardized interviews to make diagnoses but prisoners are not an easy group to survey. Many of them (and especially the men) were emotionally inarticulate to an extreme degree. They reported extreme stress but were often unable to describe their feelings with any precision. Emotion was directed outwards, in resentment towards other individuals or the institution. In some cases, there was just a desire to 'smash up' (the cell) or 'cut up' (the body), without any appreciation or understanding of the impulse.

The use of clinical criteria, by an experienced psychiatrist, is probably the best that can be achieved when attempting to make a diagnosis in such circumstances. There is no easy solution to these problems and it is certain that no standardized interview can overcome them. They should be borne in mind when interpreting the results of prison surveys. All figures for neurotic disorder are likely to represent an underestimate of the true level of psychopathology.

Leaving aside the methodological difficulties, the following are examples of the way in which the diagnosis was used in the study.

Case histories

Case no. 128
An 18-year-old woman who had served only six days of a nine-month sentence for assault. This was her first conviction, arising from a fight with a female acquaintance, over approaches to her husband. 'Things just got out of hand, she was just as much to blame. I've been on bail since the offence, I never thought I would get sent down.' She had attended normal schools but was assessed by an educational psychologist at the age of 14 because of disruptive behaviour following rows between her parents. No treatment was recommended and she was never suspended or expelled from school. At the time of her offence, she was a housewife. There was no history of drug or alcohol abuse.

At interview, she was articulate and pleasant but tearful and very distressed, terrified of prison and what she feared might happen to her there. She reported poor concentration, sleep and appetite but was not suicidal. She had never experienced symptoms like this before and attributed them to her situation. An appeal against sentence was going through the courts and she saw an early release as the best solution to her problems. Within prison, she felt desperate for someone with whom she could share her anxieties.

The researchers gave her a diagnosis of adjustment reaction (ref. no. 309.0 in ICD9), as the symptoms were so clearly related to an adverse event.

Comments
1. The experience of imprisonment, for the first time, precipitated anxiety and depression in a woman with no previous psychiatric history. This type of reaction is more common in remand prisoners, as most sentenced prisoners have already spent time on remand and have worked through their initial reactions to confinement.
2. This woman's reaction may be understandable but her distress is real. In this case, the risk of self-harm was low, but acute distress soon after being locked up is an important factor in many prison suicides. It is important that professional help is

readily available. Many inmates will provide support to others but this cannot be taken for granted, as some will exploit the fears of those new to imprisonment.

3. Counselling and support could be provided by many staff, including prison officers, probation officers, the chaplain and hospital officers/nurses, in addition to medical staff. However, the doctor remains of central importance in this process for two main reasons. First, there is a responsibility to make sure that medical services are accessible and sympathetic and, second, there is a need to screen out the more severe cases where the suicide risk is high and additional treatment is needed. One criticism of medical services in some prisons was that doctors confined their interest to major mental disorder (i.e. psychosis) and neglected services for neurotic disorders.

Case no. 458

A 20-year-old woman who had served nine months of a thirty-six month sentence for theft of electricity and arson. Together with her husband, she had accumulated electricity arrears of over £1000. After attempting to tamper with the meter, they set fire to the house in order to destroy all trace of the theft. At the time of the offence, additional stresses included the fact that she had a 2-year-old baby and was again pregnant (the second child was now eight months old and was with her on the Mother and Baby Unit). The index offences were her first convictions.

She had been brought up in a poor but close-knit family. Neurotic complaints as a child led to several child guidance attendances but no psychiatric contact as an adult. There was no history of drug or alcohol abuse. Her husband was unemployed and they had constant financial difficulties.

At interview, she reported symptoms beginning shortly after the birth of her child. They included multiple physical complaints, fatigue, poor concentration, insomnia and decreased appetite. She was easily upset and would burst into tears for no particular reason. Her mood was persistently depressed with pronounced guilt and self-blame, thinking that she would never be able to cope with the baby. Although anxious and uncertain of the future, she expressed no suicidal ideas. She was not receiving treatment and would have liked some but 'didn't know how to ask for it, or what to say'.

The researchers gave a diagnosis of neurotic depression (300.4).

Comments

1. Although imprisonment counts as an adverse life event for everyone, it is important to remember that many women's

lives are extremely problematic before they come into contact with the law. In this case, the main factors in the development of depression appear to be difficult social circumstances, poverty and the birth of a baby. Prison simply made the whole situation worse.
2. There is no evidence that psychiatric disorder played any part in the offence.
3. Depression of this type and severity is unlikely to respond to simple counselling and support and specialist psychiatric input is required.

Case no. 478

A 27-year-old women who had served four-and-a-half years of a life sentence for murder. Together with her husband, she had plotted the murder of an acquaintance for insurance money; the husband had carried out the act. She had one juvenile conviction for theft and said the murder had been her husband's idea. She had no previous psychiatric history but had taken an overdose while on remand awaiting trial.

At interview she was depressed, anxious and irritable, reporting disturbed sleep, appetite and concentration but no suicidal thoughts. She reported depression and anger throughout her sentence, believing her conviction was unjust. She had been treated with antidepressants but had stopped them because she feared dependence, although she still saw the visiting psychiatrist for supportive psychotherapy. Her depression had been worse recently, following bad news about her 'tariff' (the number of years she would have to spend in prison before she could be considered for release on licence).

The researchers gave a diagnosis of neurotic depression (300.4).

Comment

1. Depression can be seen as a reaction to circumstances and a failure to adjust to the conviction and sentence. Her present state was exacerbated by a common stressor in prisoners, bad news about tariff or parole dates. The first steps in treatment would be counselling and support, which she was already receiving from the visiting psychiatrist.
2. Her experience of medication is similar to that of many women outside prison; it had been tried but rejected because of the possibility of unwanted effects.

Case no. 1615
A 47-year-old woman who had served one year of a life sentence for murder. She had stabbed her husband in the course of a drunken argument.

Born into a large and poor family, she had suffered physical abuse from both parents and left home at the age of 13 years. She had received little schooling and was illiterate. She had three children aged 12, 14 and 17 from her 20-year marriage. Both she and her husband had been alcoholics 'for as long as I remember' but had continued to look after their children. Her husband was regularly violent towards her and this was often reciprocated. She had several psychiatric admissions in the past, due to alcoholism, depression and self-harm.

At interview, she was depressed and pessimistic, with occasional suicidal thoughts. She felt her actions had been justified by her husband's behaviour but regretted his death and missed her children.

Primary diagnosis – neurotic depression (300.4).

Secondary diagnosis – alcohol dependence (303).

Comment
1. An early life that provided a poor basis for coping with adversity was followed by a violent marriage, alcoholism and episodes of depression which preceded her imprisonment. Factors contributing to her present depression include the fact that she is beginning a life sentence, has lost her husband and children and is recovering from alcohol dependence.

Discussion

The meaning of the diagnosis of neurotic disorder will now be considered. A breakdown by type of neurosis is shown in Table 7.2. Women have a higher prevalence of all diagnoses.

Table 7.2 Sentenced prisoners given a diagnosis of neurosis: breakdown by type of disorder and gender

	Women		*Men*	
	n	%	n	%
Neurotic depression	15	6	37	2
Adjustment reaction	13	5	34	2
Anxiety state	9	4	23	1
Other	3	1	10	< 1
Total	40	16	104	6

The terms used are those contained in ICD9. It is important to make the point that these are labels used by psychiatrists but the syndromes described will be familiar to most people. For example, 'adjustment reaction' may refer to distress, in response to adverse circumstances, that is entirely understandable. It is regarded as a diagnosis only because of its severity, when it interferes with the ability to carry out everyday activities. Many 'cases' will have no history of psychiatric treatment and others will have received treatment from their GP rather than a psychiatrist.

Although the syndromes may be familiar and understandable, they are relevant to any account of the need for psychiatric services within prison. Only four of the thirteen women with a diagnosis of adjustment reaction had a history of psychiatric treatment before their present sentence, but eighteen of the twenty-seven women with other neurotic disorders reported treatment prior to their sentence. When asked about their current attitude toward treatment, only three of the forty women expressed no desire for psychiatric treatment at present and twenty-four had a strong desire for treatment. Women in prison are removed from many of the social supports that would help them cope with depression outside prison, so psychiatric services assume greater importance in dealing with neurotic disorder.

The most common treatment recommendation made by the researchers, in twenty-nine of the forty cases, was some form of outpatient/GP treatment within the prison. This was intended to encompass counselling or brief psychotherapy as well as medication. The other recommendations were no treatment in three women (as they did not want any), further assessment in four, where the diagnosis was uncertain, and a therapeutic community for two, where depression was accompanied by drug dependence. Two women were judged to require transfer to a hospital outside the prison system, as their symptoms were severe and had failed to respond to 'outpatient' treatment, including medication. Hospital transfer should not be reserved only for women with psychotic disorders; neurotic disorder can cause at least the same degree of suffering and may be associated with a significant risk of suicide, especially within prison.

Associations with offending
In contrast to some other psychiatric disorders, neurotic disorders were rarely important as a cause of imprisonment. Of course, they may be an important cause of offending and the relationship between depression and shoplifting is well known. Cases of this type

should be diverted at an earlier point in the criminal justice system and we found no evidence to suggest that they reach prison in large numbers.

Various analyses revealed only one statistical association with offending. Neurotic disorder was diagnosed in fifteen of the fifty-three women with a violent index offence (producing an odds ratio of 2.8, with a 95 per cent confidence interval from 1.4 to 5.9). The depression was probably a consequence of the offence in most of these cases. Seven of the women were serving a life sentence for murder. Depression and anxiety may be understandable reactions to either the length of sentence or the loss of the victim, who was usually someone with whom the offender had a long and ambivalent relationship.

In other cases, it is possible that the same psychological mechanisms led to both offending and neurotic symptoms, rather than the symptoms causing the offence. Rather than links with offending, it may be more useful to consider social causes of depression. A frequent concern of women in prison was whether their children were being looked after properly. This, and similar concerns, provide sufficient explanation for many cases of depression, without resorting to more complex theories.

From the viewpoint of service provision, causation is a minor issue. The important point is that depression, anxiety and other neurotic symptoms are common in the prison population, while women's normal coping mechanisms have often been removed by isolation from social contacts.

Comparison with community surveys

The extent of neurotic disorders in prisoners is not very different from that found in other groups of women and men. For example, a community survey of minor affective disorder in London (Bebbington et al., 1981) found prevalence rates of 6.1 per cent for men and 14.9 per cent for women. Reviewing similar studies, the authors note that, despite the use of different instruments, most 'report figures of the same order: 4–8% in men, and 8–15% in women'.

The results of the present study fall within these ranges. They provide no evidence to suggest that the relative prevalence of neurotic disorder in men and women serving a prison sentence differs significantly from that found in community surveys. It is not possible to draw conclusions about the relative prevalence of neurotic disorder in prisoners compared to people outside prison but our findings are certainly of the same order of magnitude as those emerging from community surveys. A definitive answer to this question would require the use of identical instruments for the prison and community samples and an adjustment to take account of the unusual age structure of the prison population, which is heavily biased towards young adults.

Neurotic disorder is the only diagnosis that was also common in women from overseas (Table 5.2). It has been argued above that these women have few social characteristics in common with other women serving a prison sentence. None had previous convictions or a history of psychiatric treatment and most were middle-class traders. This finding is in line with the suggestion that neurotic disorder can usually be seen as a reaction to circumstances, rather than having any particular association with offending.

8

Mentally disordered women in prison III: Substance abuse

Introduction

Drug and alcohol abuse would be regarded by many people as social or behavioural problems rather than psychiatric disorders. This is a rather futile argument. Substance abuse or dependence feature in all accepted psychiatric classifications and, whatever their origins, present frequently to psychiatrists. Also, both drug and alcohol abuse are associated with a variety of other psychiatric disorders, so they cannot be ignored by a comprehensive psychiatric service.

The uncertain status of substance abuse problems may have led to their neglect by psychiatric services in the past, but this trend has been reversed recently. The advent of HIV has led to a renewed interest in injectable drug use, and an expansion of treatment programmes with a wide variety of aims and orientations. Prisons have been a focus for much of this renewed interest, as they contain a large concentration of drug users. Alcohol abuse remains a neglected area.

Drug dependence

The figures given in Chapter 5 (Tables 5.4 and 5.5) combine the ICD9 diagnoses of 'drug dependence' and 'non-dependent abuse of drugs'. The descriptions given in ICD9 can be criticized as vague and over-inclusive. The following analyses use a stricter, operational definition of drug dependence. The definition is as follows:

1. Self-reported daily use of the relevant drug throughout the six months prior to the index offence.
2. Drug withdrawal symptoms on entering custody.
3. An acceptance by the interviewee that she was drug dependent at the time of the index offence.

Cannabis was excluded from this definition of drug dependence, on the grounds of its limited clinical importance. This definition ensures that those women given a diagnosis of drug dependence would probably be given such a diagnosis by any mental health professional. Doubtful cases are excluded. Therefore, the figures are, if anything, an underestimate of drug problems in prisoners.

According to these operational criteria, sixty-three women (24 per cent, with a 95 per cent confidence interval from 20 per cent to 29 per cent) and 189 men (11 per cent, 95 per cent c.i. from 9 to 12 per cent) were identified as drug dependent. Table 8.1 shows the main drugs on which these inmates were dependent. Opiates predominate and account for the higher prevalence of drug dependence in women.

Injecting during the six months prior to the index offence was reported by forty-two women and 127 men, about two-thirds of those dependent on drugs in each case. Previous treatment was reported by forty-one drug-dependent women (65 per cent) and eighty-five men (45 per cent), a statistically significant difference (the odds ratio = 2.3, with a 95 per cent confidence interval from 1.3 to 4.1). Even when the diagnosis is identical, women are more likely than men to have had psychiatric treatment.

There was marked regional variation in the prevalence of dependence, found in seventeen (27 per cent) of the sixty-three women resident in London and in nineteen (53 per cent) of the thirty-six from the Mersey health region (the odds ratio = 3.0, with a 95 per cent confidence interval from 1.3 to 7). Similar regional variation is found in male prisoners and is described in more detail elsewhere (Maden, Swinton and Gunn, 1992).

Drug-dependent women tend to be younger, their mean age being 26.4 years, compared to 29.1 years for other women in the sample (t = 2.20, p < 0.05). Despite their youth, they generally had a

Table 8.1 Drug-dependent prisoners, by gender and main type of drug

	Women N = 258		Men N = 1751	
	n	%	n	%
Opiates	48	18	117	7
Amphetamines	7	3	37	2
Cocaine	5	2	16	1
Others	3	1	19	1
Total	63	24	189	11

more extensive criminal record than women who were not drug dependent.

Some accounts of addiction in women have emphasized the link with prostitution but only three drug-dependent women said that prostitution was important in funding their habit. Women may have been reluctant to admit involvement in prostitution but prostitution-related convictions were unusual. Of the fifty-eight drug-dependent women on whom a full criminal record was available, only seven (12 per cent) had any record of convictions for offences of this type. It may be that those women who do combine addiction and prostitution do not generally find their way into the sentenced prison population, though they may spend many periods on remand.

Table 8.2 shows the index offences for drug-dependent women compared to the remainder of the sample. The drug-dependent women are mostly serving sentences for acquisitive offending. Although drug offences are also common, most drugs offenders in the sample were not drug dependent. This remains true despite the exclusion of foreign drug couriers, who would further increase the proportion of non-dependent drug offenders.

Case histories

Case no. 419

A 20-year-old woman who had served five months of a two-year sentence for the possession of heroin with intent to supply. Raised in Liverpool, in a close but poor family, she was placed in care at age 13 years after a conviction for shoplifting. She became dependent on heroin when she was 17, originally 'chasing the dragon' and later injecting. 'I used to shoplift at first but I was getting caught and I had two children to look after on my own, so I started dealing

Table 8.2 Female sentenced prisoners: index offence by drug dependence

Index offence	Non-dependent		Drug dependent		Total	
	n	%	n	%	n	%
Theft	68	35	23	37	91	35
Drugs	28	14	23	37	51	20
Burglary/robbery	30	15	11	18	41	16
Violence	48	25	5	8	53	21
Arson	11	6	–	–	11	4
Other	10	5	1	2	11	4
Total	195	76	63	24	258	100

from the flat, just to people I knew. It was two years before I was caught.'
This was her first prison sentence. 'I had a hard time in Risley, cold turkey. I told people about the drugs problem but they just put me in a strip cell when I was coming off. I've heard nothing since. I want to stay off, I've been glad for the break from gear. My two kids are with their aunt and I want them back when I go out. I was a real mess when I came in, I'd been using a gram a day and was really thin. I don't want to go back to that.'
The only diagnosis given by the researchers was drug dependence (304.0).

Comment
1. The diagnosis is uncontroversial. After several theft convictions, this woman had adopted a settled lifestyle revolving around her drug addiction and her children and there was no evidence to justify additional diagnoses.
2. Most drug users in the study reported selling drugs to other users and many justified dealing as a less stressful (and antisocial?) way of supporting their habit than theft, while recognizing that the law views supplying drugs as a more serious offence.
3. This woman's experience on remand was not untypical of drug users' accounts of withdrawal in custody. Treatment for drug users varied between different prisons and some doctors and other staff had punitive attitudes, while other prisons provided treatment of a high standard. It is now official policy that appropriate treatment is provided during withdrawal in all prisons.

Case no. 453
An 18-year-old woman who had served two months of a twelve-month sentence for burglary. Having been brought up on a large and neglected housing estate in Liverpool, she had used heroin since the age of 15 years, never injecting but 'chasing' a gram a day at the time of her arrest. Her habit was funded by burglaries, which had brought her seven court appearances and three previous custodial sentences. At interview she reported: 'All my mates are into heroin, I moved to another town to get away from it. I don't want any help. I'll stay away and stay off it. People can't help you anyway, it's something you have to do for yourself.'
The only diagnosis given by the researchers was drug dependence (304.0).

Comment
1. Similar accounts were obtained from many women. A fuller description of young people heavily involved in drugs and crime is contained in Burr's (1987) anthropological account of heroin use on a South London housing estate.
2. The attitude towards treatment was typical of inmates who did not want treatment, i.e. she was not resigned to returning to drug use but saw it as an individual problem to which treatment was not relevant.
3. A record of three custodial sentences is unusual in an 18-year-old woman. Drug users tended to have the most extensive criminal histories. The frustration of staff is understandable when a drug user refuses all offers of help, even when she appears to be locked into a cycle of drug use and repeated offending.

Case no. 415
A 35-year-old woman who had served fifteen months of a forty-month sentence for handling stolen goods and the possession of a small amount of heroin. She had first attended a drug clinic twenty years ago and claimed to have been 'the youngest registered addict in London'. She had continued to use heroin, injecting throughout this time. Although heroin was her main drug, she would also inject cocaine and amphetamine 'on most days'. A total of thirty court appearances include twenty-one convictions for prostitution-related crime, six drug and five theft offences. This was her fifth prison sentence (the previous four sentences totalled sixteen months of imprisonment). She was heavily involved in a criminal and drug-dealing subculture, financing her habit mostly by dealing.

At interview she stated: 'I inject and I never share needles. I've tried every treatment, several times. I need drugs – it would be better to stop but I'll die an addict.'

The only diagnosis given by the researchers was drug dependence (304.7).

Comment
1. This case was representative of the older heroin users in the study, resigned to continuing drug use and with extensive, unsuccessful experience of treatment. It is difficult to make any constructive suggestions.
2. Seventeen of the sixty-three drug-dependent women were over 30 at the time of their sentence. They were more likely to have started injecting heroin at an early stage, whereas younger

users started by 'chasing' (inhaling) and may not progress to injecting.

Case no. 423

A 43-year-old woman who had served two months of a thirty-six month sentence for conspiracy to supply heroin. She had grown up in a close-knit, criminal family and was married with a 15-year-old son. She had seven convictions for theft, beginning at the age of 16 years, but no previous prison sentences. She reported 'making a comfortable living from crime' along with her husband. At the age of 37 years, she became addicted to heroin, when the drug became commonplace in the criminal subculture of which she was a member.

At interview she reported: 'I tried it at a party and just liked it. My husband thinks I'm mad, he doesn't touch it. I started dealing, mostly for money, there's a lot of money in it. I never injected but I was getting through a few grams a day by the time I was caught. I just couldn't control my own habit.

'They (the court) confiscated the house and everything (because it was part of the proceeds of drug-related crime). It doesn't bother me – I can always buy another. I won't go back to drugs. I don't need treatment but I've had enough of them. What I did was wrong but it's water under the bridge. Drugs have wrecked our city though, I'd like to do something to stop kids getting involved.'

The only diagnosis given by the researchers was drug dependence (304.0).

Comment

1. This case illustrates the complexity of links between drug use and crime. This woman had grown up outside drug-using circles before her late exposure to heroin. Despite her pre-existing, extensive involvement in criminal activity, she described an escalation of offending after her introduction to heroin, the financial turnover from drugs being 'beyond anything I had ever dreamed of'.

2. It is futile to look for a simple answer to the question of whether drugs cause crime or vice versa. This drug user was a career criminal long before becoming dependent on heroin. Nevertheless, her offending escalated when she became addicted. The important message is that drug dependence is most unlikely to decrease the level of offending and, in most cases, will lead to an increase.

Discussion

Drug dependence or abuse is the commonest diagnosis in both sexes and it is twice as common in female prisoners. There are no comparable studies of drug dependence in UK prisons but American studies have found that women in prison have higher rates of drug dependence than men. However, it would be wrong to conclude from these figures that the courts are more willing to imprison drug-dependent women compared to similar men.

Estimated from the figures in Table 5.4, the sentenced prison population contains about 280 women who were dependent on drugs before entering prison, compared to about 4300 men. In other words, women account for 6 per cent of sentenced inmates who were dependent on drugs at the time of their index offence.

In the UK as a whole, the proportion of women among new addicts notified to the Home Office was 29 per cent in 1988 and had been between 26 per cent and 29 per cent throughout the preceding ten years (Home Office, 1989b). This suggests that women who are dependent on drugs are under-represented in the sentenced prison population compared to their male counterparts but the degree of under-representation is less than that of women who are not drug dependent, compared to non-dependent men (women make up less than 3 per cent of the sentenced prison population as a whole). Drug dependence appears to be an important factor contributing to the imprisonment of women. It is the most common problem that will confront mental health professionals working with female offenders.

Despite being younger than other women serving a prison sentence, those with a history of drug dependence tend to have a more extensive criminal history, but their offending is specialized, being mainly acquisitive. Drug-dependent men serving a prison sentence share these characteristics (Maden, Swinton and Gunn, 1992). They find their way into prison as a result of repeated offending, the proceeds of which are mainly spent on buying drugs.

Most drug users were from inner-city estates and funded their drug use by theft or by dealing in drugs. They resemble the women in d'Orban's description of heroin users in Holloway in 1970, although the mean age of his (mainly remand) sample was only 20 years. Some differences between the two studies are of interest. None of the drug users in d'Orban's sample faced a charge of selling drugs and most were using relatively small amounts of prescribed drugs. In the present study, practically all drugs were illicitly obtained and dealing was a common method of financing drug use, albeit one that attracted heavy sentences. Many of the dependent women would fit the description of the 'new heroin users' (Pearson, 1987), with offending preceding drug use but escalating rapidly with the onset of dependency (Jarvis and Parker, 1989).

Prostitution was unusual among the drug-dependent women in our sample but this may be a selection effect. Prostitution is not an offence and soliciting is not an imprisonable offence, so a sample of sentenced women may exclude an important group of heroin users who are able to fund their habit without risking prison.

While many women reported that their heroin use started with 'chasing the dragon', most had moved on to inject. Detailed information about injecting practices was not collected but other drug users passing through the prison system have been found to engage in high-risk behaviour (Carvell and Hart, 1990). Although the number tested is not known, two of the drug-dependent women were HIV positive.

A survey of sentenced prisoners is the wrong place to look for information about the relationship between drugs and offending but the case histories described above show a diversity in this respect that warns against any single, simple explanation of these intertwined behaviours.

It is remarkable that no statistical association was found between drug dependence and several indicators of psychological disturbance, including deliberate self-harm, a history of being in care, violence or a diagnosis of personality disorder. In general, these were normal (albeit criminal) women, who had become dependent on drugs. Burr's (1987) description of the impact of heroin on a council estate stresses the changes brought about by the arrival of a cheap, exciting and profitable commodity into a cohesive, working-class community with existing high levels of criminality. Her account emphasizes economic and social influences on the spread of drug use, regarding psychological factors as of less importance. The results of the present study are consistent with such a model. The situation is rather different for alcohol, as will be seen in the following section.

Alcohol abuse and dependence

As with drug dependence, the following account uses an operational definition of problem drinking, rather than the less specific ICD9 definitions.

Alcohol plays a circumstantial role in much offending and thirty-nine women (15 per cent) reported drinking prior to their index offence. A total of thirty-six women (14 per cent) were rated as having a significant drink problem, including ten (4 per cent) regarded as probably dependent and nine (4 per cent) definitely dependent on alcohol. A total of 374 men (21 per cent) were problem drinkers, including eighty-one (5 per cent) rated as probably dependent and thirty-four (2 per cent) definitely dependent. In contrast to drug dependence, men are more likely to be rated as problem drinkers. The male:female odds ratio is 1.7, with a 95 per cent confidence interval from 1.2 to 2.4.

At the time of sentencing, the mean age of the problem drinkers was 28.5 years, compared to 28.4 years for the rest of the sample. Unlike drug users, problem drinkers in prison do not tend to be younger than average.

Also unlike drug dependence, problem drinking showed a statistically significant association with a history of local authority care and with deliberate self-harm. Problem drinkers are significantly more likely than other women prisoners to have a record of violence, another contrast with drug users. In those women on whom CRO data was available, problem drinkers were significantly more likely to have convictions for violence against the person, arson and, predictably, drunkenness. Full details of these analyses can be found in Maden (1992).

Case histories

Case no. 409

A 27-year-old woman who had served twenty-three months of a forty-eight month sentence for robbery: while engaged in prostitution and together with male accomplices she had beaten up and robbed a client. She had been born into a family of 12 children, her father drank heavily and she was herself a heavy drinker by the age of 16 years. Her first conviction at 18 years was followed by twenty-six court appearances, including twenty-two prostitution-related convictions, numerous disorderly conduct offences and one previous conviction for robbery, in similar circumstances to the index offence, for which she had served her only previous custodial sentence.

She had been dependent on alcohol for many years but had received treatment only in prison. Outside prison, she worked as a prostitute and lived with her husband (also a heavy drinker) and their 10-year-old son.

At interview, she described herself as feeling depressed, bored and empty without drink. She had taken several overdoses in the past and used drink 'to relieve tension, to relax'. Excessive irritability caused her to get involved in many rows and fights, usually with her cohabitee; she felt they were equally to blame for these incidents. She was seeing a psychiatrist, receiving treatment with anti-depressant medication and counselling about her drinking and temper. She valued this treatment but was resigned to a return to alcohol when she left prison.

The researchers gave two diagnoses: alcohol dependence (303) and personality disorder (301.9).

Comment
1. She was given an additional diagnosis of personality disorder because of her long-standing symptoms of depression and tension. An alternative formulation would have been a chronic neurotic disorder, or simply the consequences of chronic alcoholism. The additional diagnosis was favoured because the neurotic symptoms remained prominent after almost two years without alcohol.
2. The link between alcohol and her own and her husband's violence is complex. Both her drinking and her violence could be seen as the result of her underlying anxiety, tension and difficulty in controlling her anger. She was reluctant to portray herself as a battered wife and saw her husband's problems as being similar to her own.
3. The need for help is clear. Basic counselling is needed but she would also require an alternative method of dealing with her anxiety and depression, rather than resorting to alcohol. If there is to be any prospect of success, treatment would have to continue outside prison.

Case no. 447
A 31-year-old woman who had served five months of a fifteen-month sentence for assault. She had attended normal schools but was in care from 11 to 16 years following a family breakup, when she also received child guidance. She had been drinking heavily, mainly in bouts at weekends, since the age of 14 years. Her first conviction, at 14 years, was followed by six other court appearances for petty theft and assault, all of which were drink related. The index offence was the most serious, when she stabbed a woman drinking companion during the course of an argument.

At interview, she reported a drink problem but no symptoms of alcohol dependence. She would often drink until unconscious, waking with no recollection of activities of the previous night. She had no memory of the index offence but accepted she was guilty. She reported minor neurotic symptoms, with a CIS score of eight.

She was not interested in treatment for her drink problem, stating that she could deal with it alone.

The researchers gave a diagnosis of non-dependent abuse of alcohol (305.0).

Comment
1. There is no justification for any additional psychiatric diagnosis. The link between violence and alcohol in this case may be mainly the result of acute intoxication.

Case no. 889
A 21-year-old woman who had served two months of a five-month sentence for driving with excess alcohol and driving while disqualified. Her childhood was uneventful. She started drinking at 14 years and was drunk every day by the age of 16 years, when she began to receive outpatient treatment for depression and alcoholism. She was first convicted at 17 years and had a total of five court appearances, for petty theft and drink-driving. At the time of her index offence, she was drinking every morning as soon as she woke and had the 'shakes' on entering prison, when she was treated with benzodiazepines.

At interview, she described herself as an alcoholic and reported that alcohol had led to the breakup of several relationships. She showed some neurotic symptoms, scoring ten on the CIS.

The researchers gave a diagnosis of alcohol dependence (303).

Discussion

The Epidemiological Catchment Area study of community rates of psychiatric disorder in the USA found alcohol dependence or abuse in 0.9 per cent of women and 5.0 per cent of men. Prison studies which compare women and men find less marked differences in rates of alcohol abuse, e.g. 47 per cent for women and 53 per cent for men (Guze, 1976), 4 per cent of women and 7 per cent of men (Novick, Penna and Schwarz, 1977). The present study is in line with these findings. The rate for men is higher than for women, but not dramatically so.

There are important differences between women with a drink problem and those dependent on drugs. The drug-dependent women have a very high level of acquisitive offending compared to other inmates but a lower level of violent offending. The problem drinkers have a higher level of violent offending and arson. They are also more likely to show other features of psychological disturbance, including a history of having been in care and past deliberate self-harm.

The association of alcohol with violent offending is consistent with the findings of a number of other studies, although it is recognized that the association is complex. In the present study, eight of the twelve problem drinkers with a high score for violence on the criminal profile scale were serving a sentence for murder or attempted murder; the nature of the study population means that the association is with serious violence. In his survey of fifty-nine men and seven women charged with murder, Gillies (1976) found that thirty-six (55 per cent) were affected by alcohol at the time of the offence and he described five of the thirty murderers with no other psychiatric diagnosis as 'heavy drinkers'. He noted that seventeen of the victims were also under the influence of alcohol at the time of the assault. Virkkunen found alcohol use by the perpetrator and/or the victim to be a precipitating factor in ninety-two of the 116 homicides recorded by police in Helsinki between 1963 and 1968.

The presenting findings, which refer to problem drinking in the offender rather than alcohol as a precipitating factor, suggest that the association between alcohol and violence persists in the sentenced prison population. The association of problem drinking with many other indicators of psychological disturbance is also consistent with accounts which emphasize the complexity of the links between alcohol and offending. Sentenced prisoners are not representative of all offenders. However, in the case of serious violence, the presence of a drink problem is not likely to result in diversion from the penal system, a point which will be taken up in the final chapter.

Chapter summary

Drug and alcohol abuse are the most common health problems encountered in prison populations, and drug abuse is a particular problem in women, where it is found in almost a quarter of sentenced prisoners. There is evidence to suggest that the size of the problem has increased in the last few years and, as a result of public concern, drug users are often given long sentences. Drug dependence shows an association with recidivist, acquisitive offending but not with most social and psychological characteristics measured in the survey. By contrast, alcohol problems are associated with violent offending and with indices of social and personal disturbance. Treatment facilities within prisons remain variable.

Psychiatric disorders in prisoners have now been described both in statistical terms and with the help of examples. The following chapter describes how mentally disordered women reach prison, and reviews previous work on gender differences in the processing of mentally disordered offenders by the criminal justice system. The findings of the present study are placed in the context of this work and the implications for prison health services are discussed.

9

How do mentally disordered women get to prison?

Introduction

The psychiatric problems of women in prison have been described in detail. In this chapter, the findings are placed in a wider context. How do men and women with severe psychiatric disorders get to prison? Why do some remain in prison, while others are transferred to hospital? Do gender differences arise because the system is too ready to 'psychiatrize' female offenders, or is the real problem a reluctance to accept and treat similar psychiatric disorders in men?

The answers to these questions are not to be found in sentenced prisoners but in other parts of the criminal justice system. The chapter begins with a review of gender differences in the treatment of mentally disordered offenders at various stages in the criminal process. There is a particular emphasis on the treatment of remand prisoners, as most sentenced prisoners will have been through the remand stage. The second part of the chapter considers the implications of the findings for the provision of services, both outside and inside prisons.

Gender differences in the treatment of mentally disordered offenders

Monitoring and screening procedures regulate the progress of mentally disordered offenders through the criminal justice system, with the aim of diverting many into the health service (Joseph, 1990). Diversion is formalized in the Mental Health Act 1983, which allows various forms of psychiatric treatment to be imposed by a court instead of punishment. In addition, diversion occurs in an informal way at many points in the system. For example, the relatives of a psychiatric patient may call the doctor rather than police, in the case of behaviour that would otherwise constitute an offence. The police may use their discretion in deciding not to arrest an offender who is obviously mentally disordered. Wherever these processes operate, there is the possibility of gender differences.

The amount of severe mental disorder in sentenced prisoners is one measure of the efficacy of diversion. The low number of sentenced women with a psychosis may mean that diversion is generally effective in the case of female offenders. The other findings are consistent with this possibility, in revealing an increased willingness of the system to provide treatment for women with psychiatric problems who remain within prison. When mentally disordered women were found in significant numbers, as in the case of mental handicap, there was evidence of a local failure in diversion procedures.

This explanation remains no more than a possibility, and our study alone cannot confirm or disprove it. A description of sentenced prisoners is a very crude measure of events elsewhere in the criminal justice system. A fuller explanation requires direct examination of these other parts of the system, and there are several relevant studies of gender differences in the processing of mentally disordered offenders. They focus on three institutions: the courts, hospitals, and remand prisons.

Gender differences in court

Overall, women appearing before the courts are roughly twice as likely as men to be dealt with by psychiatric means (Allen, 1987: 123–6). Even so, the proportion of psychiatric disposals is small. There has been little change since 1961, when Walker (1965) found that psychiatric disposals accounted for 0.8 per cent of all women convicted, compared to 0.5 per cent of men. In this statistic, 'psychiatric disposal' is used in its widest sense, to include probation orders with a treatment condition. The figures serve as a useful reminder that, whatever the gender differences, very few court appearances result in psychiatric treatment.

Edwards (1984), in a book based on interviews with female defendants and criminal justice professionals, described the way in which gender influences the operation of the criminal justice system at all stages. She suggested that the criminality of women was more likely to be seen as an indicator of mental disorder deserving of treatment, although comparative information on male offenders was not given in her descriptive, qualitative account. Her conclusions are consistent with the figures reported above.

Of the 286 individuals found unfit to plead in the period 1976 to 1988, 12 per cent were female, a figure which corresponds quite closely to the proportion of women (14 per cent) appearing before the crown courts in this period (Grubin, 1991). There is no evidence that women are more likely to be subject to this rare form of psychiatric disposal, which implies the presence of very severe mental disorder.

A study of magistrates' courts (Gibbens, Soothill and Pope, 1977) found that 9.3 per cent of men and 7.1 per cent of women were remanded for medical reports in Inner London, while the figures in Wessex were 4.3 per cent of men and 5.5 per cent of women. These rates are similar, but gender differences emerge from further analysis. In Inner London, 6 per cent of all male medical remands were for a sexual offence, compared to none of the female medical remands. Conversely, medical remands were imposed on 31 per cent of all women charged with violent offences against property, compared to 15 per cent of men facing a similar charge. For indictable offences, hospital orders accounted for a greater proportion of male disposals, and psychiatric probation orders a greater proportion of female disposals, although there was great variation with type of offence. The figures suggest that there are complex qualitative and numerical differences between the treatment of men and women by the courts.

It is difficult to reduce these complex differences to a simple principle. Allen (1987) shed more light on the processes involved, by examining decision-making in selected trials of female and male defendants. She chose cases where psychiatric evidence was likely to be important. Her qualitative study was based on a sample of twenty-four female and twenty-five male homicides (selected at random), eleven male 'domestic' homicides and thirty-six female and thirty-three male cases referred for psychiatric reports.

Allen concluded that there was ample scope for discretion and the operation of bias, including a sexual bias, in the way courts dealt with mentally disordered offenders. This was possible because the courts had available a wide variety of psychiatric disposals (ranging from voluntary psychiatric treatment in conjunction with a probation order, through to a hospital order with restrictions), and there was great flexibility in the rationales for their use. Most psychiatric disposals are made at the point of sentencing and this is where gender bias was most apparent, favouring a psychiatric disposal for the female offender.

Allen warned against any simple explanation in terms of a systematic tendency to psychiatrize female offenders. She found many anomalous and contradictory outcomes. For example, sentencers tended to regard female mentally disordered offenders as less deranged than their male counterparts. This should make a psychiatric disposal less likely for women, but the actual outcome was the opposite. Existing provision allows a psychiatric probation order on tenuous grounds (and irrespective of whether a diagnosis is made) while 'the fashionable preference for consensual and community based treatment actually makes it easier to offer medical treatment to those offenders who seem least disturbed' (Allen, 1987: 114). The disordered male offender, who was regarded as more disturbed, appeared to face two barriers to a psychiatric disposal. First, greater weight was attached to moral and retributive factors in male cases, favouring punishment rather than treatment. Second, there was a shortage of facilities for the chronically and dangerously disordered offender, so the court may be forced to give a prison sentence because of the lack of a bed.

Allen's work was based on a selected and non-representative sample but it is valuable because it is one of the few studies to look at how courts reach decisions on psychiatric disposal. It stresses the complexity of the forces which impinge on these decisions, and provides an alternative to simplistic accounts of the 'psychiatrization' of female offenders. Finally, Allen makes the point that the discrepancy between the treatment of female and male offenders can be seen in large part as a reluctance to recognize and treat equivalent mental disorder in male offenders.

Gender differences in hospitalization

It is a mistake to regard mentally disordered offenders as a distinct category, always separate from other psychiatric patients. With the exception of the most serious offending, chance often plays a large part in determining whether similar behaviour by mentally ill persons results in arrest and prosecution on the one hand, or compulsory hospital admission on the other. Emergency psychiatric admissions therefore provide an important context for the operation of the criminal justice system.

Compulsory admission to hospital, as an emergency, is possible under two parts of the Mental Health Act 1983. The first involves the police, who may take a mentally disordered person to hospital, rather than making an arrest, under Section 136 of the Act. The second option does not require police involvement but allows detention in hospital on the authority of a doctor and a social worker, under Section 4 of the Act. In a comparison of admissions to a London hospital under these two provisions, women comprised seventy-three (30 per cent) of the 240 'police' (Section 136) admissions and fifty (55 per cent) of the ninety-one 'civil' (Section 4) admissions (Fahy, Bermingham and Dunn, 1987), i.e. women were significantly more likely to be admitted under civil rather than police provisions.

Walker and McCabe (1973) noted that women slightly out-number men in most 'civil', compulsory admissions to psychiatric hospitals, but the situation is quite different for hospital orders made by the courts. From 1961 to 1985, women were found to account for 16.4 per cent of unrestricted and 11.5 per cent of restricted hospital orders made by the courts, while the average proportion of women among those found guilty of an indictable offence during this period was 14 per cent (Robertson, 1989). Women are therefore over-represented in unrestricted hospital orders made by the courts, but the vast majority of women who receive compulsory psychiatric treatment do so under 'civil' sections of the Mental Health Act.

In a survey of transfers of sentenced prisoners to a special hospital between 1960 and 1983, Grounds (1991) found that the ratio of women to men was 1:14, whereas the corresponding ratio in the adult sentenced prison population was 1:31. At this late stage within the criminal justice system, women are more likely than men to be admitted to a psychiatric hospital. However, the transfer of sentenced prisoners was a rare event, involving only twenty-six women (and 354 men) over a 23-year period. As this is just over one woman per year, it would be a mistake to draw far-reaching conclusions.

Remand prisons

By contrast with the transfer of sentenced prisoners, the admission of remand prisoners to hospital is not uncommon. Most sentenced prisoners with a serious mental illness will have been remanded prisoners before their trial and many will have been remanded in custody specifically for the preparation of psychiatric reports. Without considering remandees, the data on sentenced prisoners give a slanted view of reality. If researchers look at people serving a prison sentence, it is not surprising that they find mentally disordered offenders who have been rejected by psychiatric services. Those who have been accepted are, hopefully, no longer in prison (unless they are awaiting a bed). Only by looking at earlier stages of the process is it possible to know if rejection by health services is common practice, or a rare failure of care.

At the time of our survey of sentenced prisoners, no systematic data were available on remanded prisoners. This deficiency has now been remedied, and a recent series of papers describes the process of psychiatric assessment in remand prisoners.

Dell et al. (1993a) looked at all women remanded to Holloway in a six-month period, in whom psychiatric intervention was considered necessary. Out of 465 new receptions, 196 fell into this category (representing 176 women, as twenty women came in more than once). The progress of these 176 women was followed until disposal at court. They were divided into two groups, psychotic (ninety-five women) and non-psychotic.

At the time of reception, a quarter of the psychotic women were receiving outpatient psychiatric treatment and 86 per cent had a history of hospital admission. Referrals to outside psychiatrists, with the aim of obtaining a bed, were made for eighty-five of the ninety-five psychotic women (85 per cent). In the fourteen cases not referred, there were specific reasons for the decision. Often it was because of rapid improvement with treatment but, in some cases, it was because of knowledge that recent referral and assessment had not resulted in admission, so a further referral was considered pointless.

Of the eighty-one women who were referred, seven were discharged from prison before the process could be completed, leaving seventy-four available for hospital placement, of whom fifty-seven (77 per cent) were admitted to a hospital bed. For our purposes, the important group are those women who were referred but not offered a bed. In some cases, there was diagnostic uncertainty but the researchers identified sixty-four referred women, recorded by the prison psychiatrists as undoubtedly psychotic. Twelve of these women (almost one-fifth) were not offered a bed. The reasons for rejection are described in detail by the authors. Sometimes, there was disagreement about whether the woman was ill (with the visiting psychiatrist's view described as 'difficult to understand' in at least one case). More commonly, there was acceptance that illness was present, but disagreement over the degree of illness, and the need for admission.

The authors conclude that custodial remands are an ineffective, expensive and inhumane way of getting help for mentally disordered offenders (even for those offered a bed, the process involves long delays).

The same researchers looked at men in HMP Brixton, a remand prison or jail serving a large part of London (Robertson et al., 1994). The methodology was similar, an examination of all cases coming to medical attention during a five-month period in 1989. After excluding referrals for physical complaints and loss of subjects due to rapid transfer, 568 cases were analysed (comprising 547 individuals) and produced 336 psychotic men. Table 9.1

Table 9.1 Outcome of medical referral in psychotic remand prisoners, by gender (data from Dell et al., 1993 and Robertson et al., 1994)

Outcome	Women		Men	
	n	%	n	%
Not referred	14	15	80	24
Referred, not visited	11	12	63	19
Referred and rejected	13	14	96	28.5
Referred and offered a bed	57	60	96	28.5
Total	95	100	335	100

compares the outcome of medical referral for psychotic remand prisoners, by gender.

A greater proportion of women than men were referred for a second opinion, but the difference does not reach statistical significance (odds ratio = 1.8, 95 per cent confidence interval = 0.98 to 3.4). However, when psychotic inmates were referred to doctors outside the prison, the process was much more likely to end in the offer of a bed when the patient was a woman (odds ratio = 3.9, 95 per cent confidence interval = 2.3 to 6.8).

These studies were not designed to show gender differences, and the authors make it clear that they are describing medical practice at particular institutions, rather than in prisons as a whole. Nevertheless, the findings cannot easily be dismissed. The prisons chosen are both large and busy remand centres, located in the same part of the country. They occupy similar positions within the female and male prison systems respectively, and there is clear evidence of a significant difference in the way women and men are processed, when serious psychiatric illness is present.

The particular value of these studies is that they contain information about the nature of the gender difference. Dell et al. express concern that almost a fifth of referred women, with an undoubted diagnosis of psychosis, were not offered beds. There is no suggestion that too many women are being treated, and no room for complacency. The paper's complaint is that existing procedures result in too many women missing out on the treatment they need. In other words, the authors identify the problem as one of under-treatment, for various reasons. The clear implication is that the gender difference is due to an even greater degree of under-treatment in male prisoners. There is nothing in either study to suggest that the difference is due to over-treatment of women.

These studies of remand prisoners leave some questions un-answered. They consider only inmates identified by prison doctors as mentally disordered, so can make no comment on how many mentally ill prisoners go undetected. They are valuable in drawing attention to an important barrier to treatment, which lies within the health service. Mentally disordered prisoners may be highly visible as a 'prison problem' but the work shows that the true problem lies elsewhere, in the failure of health services to admit and treat many of the mentally disordered prisoners who are referred. The next section considers the implications of these findings for health services.

The implications for psychiatric services

Patients who are 'difficult to place'

The studies of Dell et al. and Robertson et al. are valuable additions to a growing literature on the difficulty of persuading health services to take on mentally disordered offenders. The problem is shared by women and men in prison. Coid's (1988) study of a male prison in South East England showed that, of the 362 mentally disordered inmates referred to the National Health Service from 1979 to 1983, about a fifth were rejected by the consultant psychiatrist responsible for their care. Bowden (1978), in an earlier study of a three-month period at Brixton, found that medical officers referred fifteen inmates with a diagnosis of paranoid schizophrenia but three were turned down by the consultant who visited to assess them.

Apart from the numbers who are rejected by the health service, there is also general agreement on the characteristics which make patients a less attractive treatment proposition. Those who are rejected are more likely to have chronic problems, whether the diagnosis is treatment-resistant psychosis, or a psychosis in association with learning difficulties or organic problems. They are also likely to be difficult patients, with a history of aggression or violence, often in association with substance abuse. Their record of compliance with treatment is often poor, usually in association with a lack of insight into the fact that they are ill. The mentally disordered women identified in our study of sentenced prisoners shared many of these characteristics.

It is not surprising that patients of this type are often rejected by psychiatric services oriented towards community care. Many of these behaviours are unacceptable and intolerable in treatment settings outside hospital, that depend on reasonable levels of co-operation and compliance. As a result, some forensic psychiatrists have concluded that community psychiatry has failed (Coid, 1994), at least in respect of this difficult minority of patients.

Of course, it is unfair to judge the success or failure of community care in its entirety on the experience of a minority of patients. The point has already been made that schizophrenic patients in prison are a tiny proportion of all people with schizophrenia, and there is no reason to assume that the needs of mentally disordered offenders are the same as the needs of other mentally ill persons. However, mentally disordered offenders have an importance out of proportion to their numbers and, as the Clunis inquiry showed (1994), society at large may well judge community care on its record of success in this area.

The Reed report: recommendations for the future

The main attempt to address the special needs of mentally disordered offenders in recent years was the joint review of health and social services for mentally disordered offenders and others requiring similar services, also known as the 'Reed' report (Department of Health and Home Office, 1992). Women were considered as a group with special needs and the report draws attention to some of these. Treatment in secure psychiatric facilities will often be disruptive to family ties, because of the distance from home. The small numbers of women in each region may lead to the centralization of services, exacerbating this problem. In addition, women are more likely to be treated at inappropriately high levels of security, for example within the special hospitals. Centralization of services means that women are present in sufficient numbers to allow provision of a range of facilities (about 20 per cent of special hospital patients are women, whereas the figure on some regional secure units may be less than 5 per cent), but the cost to the patient is confinement at a level of security which is often not justified by the risk to the public. There is a need for specialized units for female mentally disordered offenders at the level of medium security or below.

Specific provision for women is necessary, if their treatment is to take place in a safe environment. Many existing regional secure units provide an inferior service for female patients, as they contain a majority of men and, being small units, cannot provide segregated facilities. Female mentally disordered offenders have often been victims of abuse in the past, and the key to treatment may lie in the link between early victimization and later offending. This problem can only be addressed if the patient's safety is assured, both by means of the physical surroundings and by the use of appropriately trained and aware staff. Women are likely to benefit from the dawning (and overdue) realization that the care and treatment of offenders requires expertise in treating victims, and this is now being reflected in the training of those who work in forensic psychiatry.

Facilities designed for women imply a degree of segregation. A full discussion of this issue is beyond the scope of this book, but the author's view is that it is simplistic and misleading to see full segregation as either necessary or sufficient for safe and effective treatment. The Reed recommendations do not emphasize segregation, and (rightly) concentrate on the characteristics that define good services for women.

In addition to the specific needs of women, the general principles of 'Reed' are highly relevant. The central principle is that mentally disordered offenders remain the responsibility of health and social services, rather than the criminal justice system. This leads on to a consideration of the difficulty of moving patients from prison to the health service. If existing systems are to work more effectively, the problem of 'perverse incentives' must be addressed. At present, it costs the health service nothing to leave a mentally disordered person in prison, whereas adequate treatment may be a long-term enterprise, representing a considerable drain on resources. This problem is central to the predicament of mentally disordered women (and men) in prison. It will be elaborated further in the following discussion of prison health services and their relationship to the world outside.

The implications for prison health services

The clear message emerging from the findings discussed above is that the health care of prisoners would be improved by closer links between prison medical services and the National Health Service. This is not a new idea. Smith (1984) described the history of the Prison Medical Service, arguing strongly that its separation from the rest of the health services has worked against the interests of prisoners in a variety of ways. The House of Commons Social Services Committee (House of Commons, 1986) argued for a merger of the two services, but this was opposed by the then Director of the Prison Medical Service. More recently, an efficiency scrutiny (Home Office, 1990b) recommended that the Prison Medical Service become a 'Prison Health Service', whose role would be as a purchaser of services for prisoners, the main providers being NHS doctors contracted-in from outside hospitals. This proposal has now been implemented, but it is too early to judge its success.

The principle, that the same doctors who provide care outside prison should also treat prisoners, is to be welcomed. However, many practical questions remain to be addressed. For present purposes, the most important of these questions concerns funding arrangements for mentally disordered prisoners who may require transfer to outside hospitals. The findings above suggest that most severely disordered, sentenced prisoners have been identified but rejected by the NHS. Under present arrangements, a mentally disordered prisoner costs the district or region nothing, until transferred from prison to hospital. There is a financial disincentive to transfer patients and, as the internal market within the health service develops, doctors and managers will be forced to become more aware of the financial implications of transferring a prisoner. These pressures encourage delay. It remains to be seen whether the principle of 'money following the patient' can be applied to create financial incentives for an efficient system of transfer.

While mentally disordered people remain in prison, either awaiting transfer or because their disorder is not severe enough to warrant transfer, closer links with the NHS should help to improve the quality of their care. Women's prisons are already closer to the standards of the NHS in one important respect, as they employ trained nurses rather than hospital officers to provide nursing care. Male prisons are now developing in the same direction, and the present policy is to continue to increase the number of trained nurses working in prison health-care centres.

Some specific implications of the findings will now be considered, according to the type of disorder involved. For simplicity, two groups are identified. The first is serious mental disorder, encompassing psychosis, mental handicap, and severe personality disorder. This group will also include some cases of severe neurotic disorder that have failed to respond to treatment within prison. The common factor is that adequate treatment is likely to require transfer out of prison to hospital. The second group comprises substance abuse, along with less severe cases of neurotic disorder and personality disorder, when transfer to hospital is not likely to be appropriate.

Psychosis and other serious mental disorders

The findings of the present study, and the arguments presented above, suggest that serious mental disorders will be encountered in a greater proportion of female prisoners than in men. To this extent, the prevalence of severe mental disorder is seen to justify the closer psychiatric surveillance of the female prison population

described by Sim (1990). The short-term aim of psychiatric intervention is clear: to secure a smooth and rapid transfer to hospital.
It was apparent from our results that this was not achieved in all
cases. Despite the closer psychiatric scrutiny that already exists,
some forms of severe mental disorder remained significantly more
common in female prisoners, and a small number of women were
being held in inhumane and unsuitable conditions, when they
should have been in hospital. It was argued above that this reflected
the peculiarities of one institution, but it emphasizes the fragility of
those mechanisms which limit the number of mentally ill women in
prison. Given the reluctance of many parts of the health service to
accept difficult patients, irrespective of gender, it becomes doubly
important that doctors working in prison explore all possibilities
for transfer. In recent years, there has been an encouraging increase
in the use of those parts of the Mental Health Act which permit
the emergency transfer of prisoners to hospital for psychiatric
treatment.

As Table 9.2 shows, the number of requests for transfer almost
doubled in the four years from 1988 to 1992, and there was a
corresponding increase in the acceptance of patients (HM Prison
Service, 1993). This provision was previously under-used, but has
the advantage of allowing ready movement of mentally ill prisoners,
without reference to the courts. These figures represent a major
improvement in the health care of prisoners, yet they required no
new legislation or structural changes, just a change in attitudes and
working practices. This change was driven by doctors working in
the prisons, and it is a good illustration of their power within the
system, when they choose to exercise it.

In addition to organizational and practical changes, there remain
questions which the medical or psychiatric profession must address.
One concerns the quality of reports and the maintenance of
adequate professional standards. A doctor's actions in preparing a
report may have dramatic and long-lasting consequences for the
offender-patient, yet the process of obtaining reports seems
haphazard. A first step in establishing standards would be to
insist that psychiatric reports are prepared by a doctor with a

Table 9.2 Number of prisoners for whom transfer to hospital was requested and achieved under S47 and S48 of the Mental Health Act, between 1988 and 1992

	1988/89	1989/90	1990/91	1991/92
Number of requests	224	250	326	415
Number for whom transfer was approved	184	220	323	379

psychiatric qualification, which is not always the case under existing arrangements.

If this measure were adopted, the profession would still have a responsibility to ensure adequate training in the preparation of reports, and audit of their quality. This responsibility is not taken seriously at the moment. During the course of the prison study, the researchers saw many reports that contained strong elements of censure and calls for punishment, including active opposition (on moral grounds) to programmes of treatment offered by other doctors. Legislation giving patients access to their health records came into force in November 1991, and applies equally to medical records within prisons. This may moderate some of the worst excesses, and will presumably open the way for offender-patients to sue doctors whose reports fail to meet reasonable standards. The Royal College of Psychiatrists should also address the training and ethical issues involved in this area, where an individual doctor can exercise enormous power with very limited accountability. Psychiatric reports provide ideal material for medical audit, as they summarize an assessment within a short document.

The problem of inadequate or unfair reports is not confined to prison doctors, and some of the worst examples were prepared by NHS doctors justifying decisions not to take patients. However, the problem is exacerbated within prison, by the professional isolation of prison doctors and the lack of a well-defined career structure (Smith, 1984: 94–101). The development of a recognized training programme and a postgraduate qualification in prison medicine would address this problem, and make prison medicine a more attractive career choice.

Many accounts of prison psychiatry dismiss its role in the management of female prisoners as entirely negative. One of the messages of this book is that the effects of prison psychiatry may be good or bad, largely dependent on the way in which it is practised. This is never more so than in the case of prisoners requiring a move to hospital.

Personality disorders, substance abuse and neurotic disorders

For most patients with these diagnoses, transfer out of prison is not appropriate. Even if diversion is very efficient, prison populations will contain large numbers of women with these diagnoses, so the development of appropriate services within prisons is essential.

The high prevalence of substance abuse and personality disorders in prison populations has long been apparent, and there have been many calls for improved services. The response has been variable and uncoordinated, often depending on the interests of particular officers or probation staff. In this setting, it may seem pointless to restate the need for services. However, there are signs of recent improvements, due to pressure from three different directions.

First, the advent of HIV has revived interest in treatment of the addictions. At the same time, a recognition that drug abuse and offending are intimately linked has increased interest within prisons. The second important factor was the Strangeways riot (Woolf and Tumim, 1991), which drew attention to both prison regimes and the problem of difficult prisoners. The system had become resigned to doing nothing more than 'humane containment', whereas outsiders looking in at prisons are asking for more attention to rehabilitation. The third factor is the granting of agency status to the prison service, and the appointment of a Director General from outside the service, with a background in industry. This initiative goes along with privatization of parts of the system, which may prove unsuccessful, but the positive aspect is bringing outside attention in to a prison system that can easily become inward looking, demoralized and negative.

The time is therefore right for new treatment initiatives within prison and some developments are already in operation. For drug users, there is now a policy of throughcare, based on the principles of education and self-help, and ensuring that prisoners are given the opportunity to establish links with outside agencies. The intention is that these links will be maintained after release. Holloway prison has been central to this development, by encouraging drug treatment agencies to come into the prison and take part in programmes. As drug dependence is the single most common problem in female prisoners, it is appropriate to develop programmes in women's prisons, before attempting to apply them to the rest of the system.

HMP Holloway already operates a withdrawal regime for drug-dependent inmates and efforts are being made to extend this approach to more (female and male) prisons. It is difficult for the prison service to do more than produce guidelines, as the decision to prescribe depends ultimately on individual doctors. This is yet another reminder of the importance of proper education, training and supervision for doctors working in prisons, as a policy can only be as good as the individuals implementing it. The situation has been improved in some prisons by the employment, on a sessional basis, of an addictions specialist from outside the prison.

By contrast with dependence on other drugs, alcoholism remains neglected. Many of the principles of treatment are the same, and it may be that the initiatives within addictions will generalize.

It is difficult to make any rigid distinction between the problems of substance abuse and personality disorder, as they often coexist, but a difference of emphasis is appropriate. In HMP Grendon, the prison system has provided a model for the treatment of personality disorder, in a therapeutic community, which is unmatched outside prison except in small parts of the health service (e.g. the Henderson and Cassell hospitals, and some wards in the special hospitals). Grendon is described in Gunn et al. (1978) and, more recently, in Genders and Player (1989). It will not be discussed in detail here, as it is available only to male prisoners.

There is no therapeutic community available to women within the prison system. It is difficult to justify this state of affairs, as personality disorder is more common in female prisoners. The small size of the female prison population may preclude the use of a full prison for the purpose, but it should be possible to develop one wing of a prison to provide a regime of this type.

However, a therapeutic community is not the only response to personality disorder, nor should it be the main response. It is an unsatisfactory diagnostic label that encompasses a range of social and personal problems, requiring a range of solutions. Past experience of abuse and recent experience of unsatisfactory relationships or living conditions are common. Interventions may involve social work help, individual counselling or psychotherapy, psychological treatments and, in some cases, medication. Much treatment of this type is already provided in prisons by prison officers, probation officers, chaplains and psychologists, in addition to the medical services. There is some scope for improvement in coordination but the main need is probably for supervision. During the course of the research, the interviewers were frequently asked for advice by personal officers who felt overwhelmed by some of the problems confronting them.

The same principles apply to the treatment of neurotic disorders, which also require a range of supports and interventions. It is worth noting that many of the criticisms of treatment within prison could also be directed at treatment outside prison. For example, general practitioners have been criticized for their over-reliance on prescribing as a response to anxiety and depression, and there is a growing demand for psychotherapy in various forms.

Limits on treatment within prisons

The treatment of addictions and personality disorder raises important questions about the limits of treatment within prison. It has already been established that the safe and effective treatment of psychosis and other serious mental illness is usually not possible within prison, but this should not be taken to imply that the treatment of other disorders is straightforward. It is easy to criticize the prisons for not doing enough, without understanding the constraints under which they operate.

Women are imprisoned as punishment, the deprivation of their liberty. They retain the freedom to refuse treatment that is offered to them. Many of the women interviewed in the present study believed their problems were behind them, and they had no interest in treatment. Once treatment has been offered, in the best possible setting, it is difficult to see what more can be done. In a prison setting, there is a fine line between encouraging people to have treatment, and coercing them. For example, parole reports may take account of a refusal to join an addictions treatment programme within the prison. The law in England and Wales does not allow compulsory treatment for substance abuse or minor degrees of psychiatric disorder and prisons should err on the side of caution in their interpretation of what constitutes compulsory treatment.

More practical difficulties include the problem of ensuring confidentiality, which may sometimes conflict with the prison's first priority, which is security. There is also competition for resources, so that money spent on treatment facilities for the mentally disordered within prisons is not available for more general improvements in the prison environment and regime. Prisons are frequently criticized for the poverty of their regimes and the restricted range of activities and training made available to inmates. It is difficult to argue for the diversion of money away from these areas into therapeutic facilities for a small number of prisoners.

The final problem is a more general, philosophical one. How much can society expect the prisons to do for those with personality or substance abuse problems, relative to what is done outside of prisons? It is easy to criticize prisons for not doing more, as personality and drug problems are concentrated and visible within them. However, the idea that such problems are found only or mainly in prisoners is, of course, an illusion. Most offenders spend more time outside prison than in it. It is unreasonable to focus only on help that could be provided within prison, while neglecting the lack of facilities outside.

Conclusion

An obvious criticism of this final chapter is that many of the points discussed apply equally to both men and women within the criminal justice system. This emphasis is deliberate. All mentally disordered offenders are disadvantaged by their status as offenders, and as a result of mental disorder. In addition, they are likely to be poor and socially disadvantaged, and a sizable minority suffer discrimination on ethnic grounds. For women, discrimination based on gender is added to this list. However, it has been shown that the influence of gender may be positive or negative, often in an unpredictable way. The additional disadvantage due to gender may be almost insignificant when set beside the disadvantages of being a mentally disordered offender, and lacking economic and social power. The impression gained during the course of this study was that female mentally disordered offenders had more in common with similar men than they did with other women who were not mentally disordered offenders. A telling and tragic example is provided by the fact that none of the women with serious mental illness had care of their own children, even before entering the criminal justice system. Debates about child care may be important for many female offenders, but had little relevance for this small group with special needs.

It remains true that mentally ill women in prison have special needs (not least, protection from abusive men), but the main problem may be the overall lack of services for mentally disordered offenders. For example, it is generally accepted that women require specialized facilities within regional secure units. The secure units generally acknowledge this need, but it is difficult to respond when the whole system is overloaded, with a relentless pressure to keep bed occupancy at a maximum. Women should therefore benefit from general improvements in services for mentally disordered offenders, such as those recommended in the Reed report, as well as from changes specifically targeted at women.

The final comments should be addressed to the main purpose of the study, the need to get severely mentally disordered prisoners out of prison and into hospital. The characteristics of the women and men concerned are similar, and tend to make them difficult to place in the health service, but there is one important difference. The number of women involved is so small that their transfer from prison to hospital is an achievable target, even without any major expansion of facilities. The existing system needs to work more effectively and efficiently. Its continuing failure to do so is an indictment of psychiatric services both inside and outside prison.

This is intended as a strong and deserved criticism of psychiatry in its dealings with mentally disordered offenders. It is not an endorsement of the traditional anti-psychiatry stance, with its blanket opposition to the psychiatrization of female offenders.

After three years of close contact with prisons, prisoners and prison doctors, it became abundantly clear that psychiatry is not a single entity. Rather than deciding for or against psychiatry in general, there is a more pressing choice. On the one hand, there is psychiatry practised well, with the patient's interests to the fore. On the other, there is psychiatry practised badly. The survey showed both extremes, with most examples falling somewhere in the middle ground. There are recent signs of general improvement, as a result of the increased attention given to the plight of mentally disordered offenders. It will be important to translate this interest into resources, to develop a range of services for mentally disordered women in prison. This task has bypassed the anti-psychiatrists and the debate on psychiatrization, which seems dated and irrelevant.

References

Allen, H. (1987). *Justice Unbalanced. Gender, Psychiatry and Judicial Decisions.* Milton Keynes: Open University Press

American Psychiatric Association (1980). *Diagnostic and Statistical Manual of Mental Disorders*, 3rd edn. Washington: American Psychiatric Association

Bebbington, P. E. and Hill, P. D. (1985). *A Manual of Practical Psychiatry.* London: Blackwell, 38

Bebbington, P., Hurry, J., Tennant, C., Sturt, E. and Wing, J. K. (1981). Epidemiology of mental disorders in Camberwell. *Psychological Medicine*, **11**: 561–79

Benn, M. (1983). Women in prison . . . breaking the silence. *Spare Rib*, November: 51–5

Bluglass, R. (1966). *A psychiatric study of Scottish convicted prisoners.* University of St. Andrews MD thesis

Bowden, P. (1978). Men remanded into custody for medical reports: the selection for treatment. *British Journal of Psychiatry*, **113**: 320–31

Box, S. and Hale, C. (1983) Liberation and female criminality in England and Wales. *British Journal of Criminology*, **23**: 35–49

Burke, A. W. (1984). Racism and psychological disturbance among West Indians in Britain. *International Journal of Social Psychiatry*, **30**: 50–68

Burr, A. (1987). Chasing the dragon. *British Journal of Criminology*, **27**: 333–57

Campbell, L. (1990). Impairments, disabilities and handicap: assessment for court. In Bluglass, R. and Bowden, P. (eds), *Principles and Practice of Forensic Psychiatry.* London: Churchill Livingstone, 419–24

Carlen, P. (1983). *Women's Imprisonment. A Study in Social Control.* London: Routledge & Kegan Paul

Carlen, P. (1985). Law, psychiatry and women's imprisonment. A sociological view. *British Journal of Psychiatry*, **146**: 618–21

Carvell, A. L. M. and Hart, G. J. (1990). Risk behaviours for HIV infection among drug users in prison. *British Medical Journal*, **300**: 1383–4

Cheadle, J. and Ditchfield, J. (1982) *Sentenced Mentally Ill Offenders.* London: Home Office Research and Planning Unit

Coid, J. W. (1984). How many psychiatric patients in prison? *British Journal of Psychiatry*, **145**: 78–86

Coid, J. W. (1988). Mentally abnormal prisoners on remand. *British Medical Journal*, **296**: 1779–84

Coid, J. W. (1991). 'Difficult to place' psychiatric patients: The game of pass the parcel must stop. *British Medical Journal*, **302**: 603–4

Coid, J. W. (1994). Failure in community care: psychiatry's dilemma. *British Medical Journal*, **308**: 805–6

Cookson, H. M. (1977) A survey of self-injury in a closed prison for women. *British Journal of Criminology*, **17**: 332–46

Cope, R. (1989). Psychiatry, ethnicity and crime. In Bluglass, R. and Bowden, P. (eds), *Principles and Practice of Forensic Psychiatry*. London: Churchill Livingstone, 849–61

Corbett, J. A. (1979). Psychiatric morbidity and mental retardation. In James, F. E. and Snaith R. P. (eds), *Psychiatric Illness and Mental Handicap*. London: Gaskell, 11–25

Daniel, A. D., Robins, A. J., Reid, J. C. and Wilfley, M. A. (1988). Lifetime and six-month prevalence of psychiatric disorders among sentenced female offenders. *Bulletin of American Academy of Psychiatry and the Law*, **16**: 333–42

Dell, S. (1991). Book review. *Journal of Forensic Psychiatry*, **2**: 225–8

Dell, S., Robertson, G., James, K. and Grounds, A. (1993a). Remands and psychiatric assessments in Holloway prison I: The psychotic population. *British Journal of Psychiatry*, **163**: 634–9

Dell, S., Robertson, G., James, K. and Grounds, A. (1993b). Remands and psychiatric assessments in Holloway prison II: The non-psychotic population. *British Journal of Psychiatry*, **163**: 640–4

Department of Health (1990). *On the State of the Public Health for the Year 1989 – The Annual Report of the Chief Medical Officer*. London: HMSO, 54–61

Department of Health and Home Office (1992). *Review of Health and Social Services for Mentally Disordered Offenders and Others Requiring Similar Services*. Final Summary Report, Cmnd. 2088

Dobash, R. P., Dobash R. E. and Gutteridge, S. (1986). *The Imprisonment of Women*. Oxford: Blackwell

Dooley, E. (1990). Prison suicide in England and Wales 1972–87. *British Journal of Psychiatry*, **156**: 40–5

Edwards, S. S. M. (1984). *Women on Trial*. Manchester: Manchester University Press

Epps, P. (1951). A preliminary survey of 300 female delinquents in Borstal institutions. *British Journal of Delinquency*, **1**: 187–97

Epps, P. (1954). A further survey of female delinquents undergoing Borstal training. *British Journal of Delinquency*, **4**: 270–1

Epps, P. and Parnell, R. W. (1952). Physique and temperament of women delinquents compared with women undergraduates. *British Journal of Medical Psychology*, **25**: 249–55

Fahy, T. A. (1989). The police as a referral agency for psychiatric emergencies – a review. *Medicine, Science and the Law*, **29**: 315–22

Fahy, T. A., Bermingham, D. and Dunn, J. (1987). Police admissions to psychiatric hospitals: a challenge to community psychiatry? *Medicine, Science and the Law*, **27**: 263–8

Farrington, D. P. (1990). Implications of criminal career research for the prevention of offending. *Journal of Adolescence*, **13**: 113.

Foucault, M. (1967). *Madness and Civilization*. London: Tavistock

Foucault, M. (1979). *Discipline and Punish. The Birth of the Prison*. London: Peregrine

Genders, E. and Player, E. (1995). *Grendon: A Study of a Therapeutic Prison*. Oxford: Clarendon Press.

Gibbens, T. C. N. (1971). Female offenders. *British Journal of Hospital Medicine*, 279–86

Gibbens, T. C. N., Soothill, K. L. and Pope, P. J. (1977). *Medical Remands in the Criminal Court*. Oxford: Oxford University Press

Gibson, E. (1975). *Homicide in England and Wales 1967–1971*. Home Office Research Study No. 31. London: HMSO

Gibson, E. and Klein, S. (1969). *Murder 1957–1968*. Home Office Research Study No. 3. London: HMSO

Gillies, H. (1976). Homicide in the west of Scotland. *British Journal of Psychiatry*, **128:** 105–27

Glick, R. and Neto, V. (1977). *National Study of Women's Correctional Programs.* Washington: Government Printing Office

Goldberg, D. P., Cooper, B., Eastwood, M. R., Kedward, H. B. and Shepherd, M. (1970). A standardized psychiatric interview for use in community surveys. *British Journal of Preventive and Social Medicine*, **24:** 18–23

Green, P. (1991). *Drug Couriers.* London: Howard League for Penal Reform

Grounds, A. (1991). The transfer of sentenced prisoners to hospital 1960–1983. A study in one special hospital. *British Journal of Criminology*, **31:** 54–71

Grubin, D. H. (1991). Unfit to plead in England and Wales 1976–1988: a survey. *British Journal of Psychiatry*, **158:** 540–8

Gunn, J., Maden, A. and Swinton, M. (1991). *Mentally Disordered Prisoners.* London: Home Office

Gunn, J., Maden, A. and Swinton, M. (1991). Treatment needs for prisoners with psychiatric disorders. *British Medical Journal* **303:** 338–41

Gunn, J. and Robertson, G. (1976). Drawing a criminal profile. *British Journal of Criminology*, **16:** 156–60

Gunn, J., Robertson, G., Dell, S. and Way, C. (1978). *Psychiatric Aspects of Imprisonment.* London: Academic Press

Guze, S. B. (1976). *Criminality and Psychiatric Disorders.* New York: Oxford University Press

Harrison, G., Owens, D., Holton, A., Neilson, D. and Boot, D. (1988). A prospective study of severe mental disorder in Afro-Caribbean patients. *Psychological Medicine*, **18:** 643–57

Hartmann, M. (1977). *Victorian Murderesses. A True History of Thirteen Respectable French and English Women Accused of Unspeakable Crimes.* London: Robson Books

Hedderman, C. and Hough, M. (1994). Does the criminal justice system treat men and women differently? *Home Office Research and Statistics Department, Research Findings No. 10.* London: Home Office

HM Prison Service (1993). *Report of the Director of Health Care for Prisoners 1992/1993.* London: HM Prison Service

Hirschi, T. and Hindelang, M. J. (1977). Intelligence and delinquency: a revisionist review. *American Sociological Review*, **42:** 571–87

Hitch, P. J. and Clegg, P. (1980). Modes of referral of overseas immigrant and native-born first admissions to psychiatric hospitals. *Social Sciences and Medicine*, **14A:** 369–74

Home Office (1986). The ethnic origin of prisoners: the prison population on January 30th 1985 and persons received, July 1984–March 1985. *Home Office Statistical Bulletin 17/86.* London: Government Statistical Service

Home Office (1989a). *Prison Statistics: England and Wales 1988.* London: HMSO

Home Office (1989b). Statistics of drug addicts notified to the Home Office, United Kingdom, 1988. *Home Office Statistical Bulletin 13/89.* London: Government Statistical Service

Home Office (1990a). *Criminal Statistics, England and Wales.* London: HMSO

Home Office (1990b). *Report on an Efficiency Scrutiny of the Prison Medical Service.* London: Home Office

House of Commons (1986). *Third Report from the Social Services Committee, Session 1985–1986, Prison Medical Service.* London: HMSO

Hunter, H. (1979). Forensic psychiatry and mental handicap. In James, F. E. and Snaith, R. P. (eds), *Psychiatric Illness and Mental Handicap.* London: Gaskell, 131–46

Jarvis, G. and Parker, H. (1989). Young heroin users and crime. How do the new users finance their habits? *British Journal of Criminology*, **29**: 175–85

Jones, A. (1991). *Women Who Kill*. London: Victor Gollancz

Joseph, P. (1990). Mentally disordered offenders: diversion from the criminal justice system. *Journal of Forensic Psychiatry*, **1**: 133–8

Kempe, C. H., Silverman, F. N., Steele, B. S., Droegemueller, W. and Silver, H. K. (1962). The battered child syndrome. *Journal of the American Medical Association*, **181**: 17–24

King, R. D. and McDermott, K. (1989). British prisons 1970–1987. The ever-deepening crisis. *British Journal of Criminology*, **2**: 107–28

Kozuba-Kozubska, J. and Turrel, D. (1978). Problems of dealing with girls. *Prison Service Journal*, **29**: 4

Larkin, E., Murtagh, S. and Jones, S. (1988). A preliminary study of violent incidents in a special hospital. *British Journal of Psychiatry*, **153**: 226–31

Lewis, A. (1974). Psychopathic personality: a most elusive category. *Psychological Medicine*, **4**: 133–40

Liebling, A. (1992). *Suicides in Prison*. London: Routledge

Littlewood, R. and Lipsedge, M. (1981). Acute psychotic reactions in Caribbean patients. *Psychological Medicine*, **11**: 289–302

Maden, A. (1992). *Psychiatric Disorder in Women Serving a Prison Sentence*. University of London MD thesis

Maden, A., Swinton, M. and Gunn, J. (1992). A survey of pre-arrest drug use in sentenced prisoners. *British Journal of Addiction*, **87**: 27–33

Maden, A. and Gunn, J. (1993). When does a prisoner become a patient? Editorial, *Criminal Behaviour and Mental Health*, **3**: iii–viii

Mandaraka-Sheppard, A. (1986). *The Dynamics of Aggression in Women's Prisons in England*. Aldershot: Gower

Maudsley, H. (1874). *Responsibility in Mental Disease*. London: King. Quoted in Lewis, A. (1974). Psychopathic personality: a most elusive category. *Psychological Medicine*, **4**: 133–40

Mayhew, H. and Binny, J. (1862). *The Criminal Prisons of London and Scenes of Prison Life*. London: Griffin, Bihn & Co

Mayhew, P., Elliott, D. and Dowds, L. (1989). *The 1988 British Crime Survey*. Home Office Research Study No. 111. London: HMSO

Menninger, K. (1969). *The Crime of Punishment*. New York: Viking

Moorehead, C. (1985). The strange events at Holloway. *New Society*, **72**: 40–2

Morris, A. (1987). *Women, Crime and Criminal Justice*. Oxford: Blackwell

Nadel, J. (1993). *Sara Thornton*. London: Gollancz

Novick, L. F., Penna, R. D. and Schwarz, M. S. (1977) Health care status of the New York Prison population. *Medical Care*, **15**: 205–16

d'Orban, P. T. (1970). Heroin dependence and delinquency in women – a study of heroin addicts in Holloway prison. *British Journal of Addictions*, **65**: 67–78

d'Orban, P. T. (1971). Social and psychiatric aspects of female crime. *Medicine, Science and the Law*, **11**: 104–16

d'Orban, P. T. (1972). Baby stealing. *British Medical Journal*, **2**: 635–9

d'Orban, P. T. (1973). Female narcotic addicts: a follow-up study of criminal and addiction careers. *British Medical Journal*, **4**: 345–7

d'Orban, P. T. (1979). Women who kill their children. *British Journal of Psychiatry*, **134**: 560–71

d'Orban, P. T. (1985). Psychiatric aspects of contempt of court among women. *Psychological Medicine*, **15**: 597–607

Pearson, G. (1987). *The New Heroin Users*. Oxford: Blackwell

Penrose, L. S. (1939). Mental disease and crime: outline of a comparative study of European statistics. *British Journal of Medical Psychology*, **18**: 39–41

Pollak, O. (1950). *The Criminality of Women*. New York: University of Pennsylvania Press

Prison Department (1987). *Report on the Work of the Prison Department 1985–86*. London: HMSO, 93

Quinton, R. (1910). *Crime and Criminals*. London: Longmans, Green & Co.

Raven, J. C. (1943). *Guide to Using the Mill Hill Vocabulary Scale*. London: Lewis

Raven, J. C. (1958). *Guide to Using the Standard Progressive Matrices*. London: Lewis

Regier, D. A., Boyd, J. H., Burke, J. D., Rae, D. S., Myers, J. K., Kramer, M., Robins, L. N., George, L. K., Karno, M. and Locke, B. Z. (1988). One-month prevalence of mental disorders in the United States. Based on five epidemiologic catchment area sites. *Archives of General Psychiatry*, **45**: 977-86

Report of the inquiry into the care and treatment of Christopher Clunis (1994). London: HMSO

Robertson, G. (1982). The 1959 Mental Health Act of England and Wales: changes in the use of its criminal provisions. In Gunn, J. and Farrington, D. P. (eds), *Abnormal Offenders, Delinquency and the Criminal Justice System*. London: Wiley, 245–68

Robertson, G. (1989). The restricted hospital order. *Psychiatric Bulletin*, **13**: 4–11

Robertson, G. (1992). *The Role of Police Surgeons. Royal Commission on Criminal Justice Research Study No. 6*. London: HMSO

Robertson, G., Dell, S., James, K. and Grounds, A. (1994). Psychotic men remanded in custody to Brixton Prison. *British Journal of Psychiatry*, **164**: 55–61

Robins, L. N., Helzer, J. E., Croughan, J., Williams, J. B. W. and Spitzer, R. L. (1979). *The National Institute of Mental Health Diagnostic Interview Schedule*. Rockville: NIMH

Roper, W. F. (1950). A comparative study of the Wakefield Prison population in 1948, parts I & II. *British Journal of Delinquency*, **1**: 15–28 and 243–70

Schonnell, F. J. (1961). *The Psychology and Teaching of Reading*. London: Oliver and Boyd

Sim, J. (1990). *Medical Power in Prisons. The Prison Medical Service in England 1774–1989*. Milton Keynes: Open University Press

Smith, R. (1984). *Prison Health Care*. London: British Medical Association

Taylor, P. J. and Gunn, J. (1984). Violence and psychosis 1. Risk of violence among psychotic men. *British Medical Journal*, **288**: 1945–9

Thornicroft, G. (1990). Cannabis and psychosis. *British Journal of Psychiatry*, **157**: 25–33

Tizard, J. (1964). *Community Services for the Mentally Handicapped*. Oxford: Oxford University Press

Turner, T. S. and Tofler, D. S. (1986). Indicators of psychiatric disorder among women admitted to prison. *British Medical Journal*, **292**: 651–3

Virkkunen, M. (1974). Alcohol as a factor precipitating aggression and conflict behaviour leading to homicide. *British Journal of Addictions*, **69**: 149–54

Walker, M. A. (1987). The ethnic origin of prisoners. *British Journal of Criminology*, **27**: 202–6

Walker, N. (1965) *Crime and Punishment in Britain*. Edinburgh: Edinburgh University Press

Walker, N. and McCabe, S. (1973). *Crime and Insanity in England. Vol. 2: New Solutions and New Problems*. Edinburgh: Edinburgh University Press

Weller, M. P. I. and Weller, B. G. A. (1988). Crime and mental illness. *Medicine, Science and the Law*, **28**: 38–53

Wessely, S., Castle, D., Der, G. and Murray, R. (1991). Schizophrenia and Afro-Caribbeans. A case-control study. *British Journal of Psychiatry,* **159:** 795–801

West, D. J. and Farrington, D. P. (1973). *Who Becomes Delinquent?* London: Heinemann

Wilkins, J. and Coid, J. (1991). Self-mutilation in female remanded prisoners I: An indicator of severe psychopathology. *Criminal Behaviour and Mental Health,* **1:** 247–67

Wing, J. K., Cooper, J. E. and Sartorius, B. (1974). *The Measurement and Classification of Psychiatric Symptoms.* London: Cambridge University Press

Wood, P. H. N. (1980). *International Classification of Impairments, Disabilities and Handicaps.* Geneva: World Health Organization

Woodside, M. (1961). Women drinkers admitted to Holloway Prison during February 1960. *British Journal of Criminology,* **1:** 221–35

Woodside, M. (1962). Instability in women prisoners. *Lancet,* **2:** 928–30

Woolf, Lord Justice and Tumim, S. (1991). *Prison Disturbances, April 1990.* Cmnd. 1456. London: HMSO

World Health Organization (1978). *Mental Disorders: Glossary and Guide to their Classification in Accordance with the Ninth Revision of the International Classification of Diseases.* Geneva: World Health Organization

Index

Note: Page references to figures and tables are in italic